CHINA'S TRADE POLICY ON INTERNATIONAL AIR TRANSPORT

This book is a political–economic analysis of China's transformation to become a global aviation power. It aims to identify the driving forces that have shaped China's ever-evolving international air transport policy direction and goals in the past four decades and further determines how and to what extent these driving forces have shaped China's considerations and strategies when executing its policy goals through bilateral air services negotiations.

The findings reveal that China's international air transport policymaking has remained in the domain of the country's aviation regulator, which has enjoyed an exclusivity to exercise its power on the air transport sector. The book argues that China's international air transport policy direction is in alignment with the country's overall strategic mission and its goal is set to support the country's endeavour to realise the "China dream." It concludes that factors at all levels interact with each other with far-reaching impact on the country's policy direction and goal setting; however, these factors are constrained by time and circumstances.

The book is a must-read for a wide array of audiences, including, but not limited to, scholars and industry professionals who have an interest in China's political economy, policymaking, international trade, government behaviour, corporate political activities, air transport, aviation liberalisation, and bilateral negotiations.

Chrystal Zhang, PhD, is an Associate Professor of Aviation at RMIT University, Australia. She has developed her expertise in aviation through her three decades of experience working with government agencies, commercial organisations, and higher education institutions in China, the UK, and Australia.

Kareem Yarde is a researcher in Australia. He has established himself as a highly proficient aviation/tourism research consultant in the Caribbean region where he has worked on projects for the Barbados tourist board and regionally. He has served as a senior examining committee member for tourism assessments in the region.

CHINA'S TRADE POLICY ON INTERNATIONAL AIR TRANSPORT

Policy Goals, Driving Forces, and Impact

Chrystal Zhang and Kareem Yarde

LONDON AND NEW YORK

First published 2021
by Routledge
2 Park Square, Milton Park, Abingdon, Oxon OX14 4RN

and by Routledge
52 Vanderbilt Avenue, New York, NY 10017

Routledge is an imprint of the Taylor & Francis Group, an informa business

© 2021 Chrystal Zhang and Kareem Yarde

British Library Cataloging-in-Publication Data
A catalogue record for this book is available from the British Library

Library of Congress Cataloging-in-Publication Data
Names: Zhang, Chrystal, author. | Yarde, Kareem, author.
Title: China's trade policy on international air transport: policy goals, driving forces and impact / Chrystal Zhang and Kareem Yarde.
Description: Milton Park, Abingdon, Oxon; New York, NY: Routledge, 2021. | Includes bibliographical references and index.
Identifiers: LCCN 2020015176 (print) | LCCN 2020015177 (ebook)
Subjects: LCSH: Aeronautics, Commercial—Government policy—China. | China—Economic policy—1949-
Classification: LCC HE9878.A35 Z43 2021 (print) | LCC HE9878.A35 (ebook) | DDC 387.7/10951—dc23
LC record available at https://lccn.loc.gov/2020015176
LC ebook record available at https://lccn.loc.gov/2020015177

ISBN: 978-1-4094-5144-0 (hbk)
ISBN: 978-0-367-53476-9 (pbk)
ISBN: 978-1-315-54927-9 (ebk)

Typeset in Bembo
by codeMantra

The book is for Luda, the most beloved spirit.

CONTENTS

FIGURES

TABLES

PREFACE

Discussions surrounding contemporary 21st century developments are often punctuated by one common thread; that is, the absolute importance of understanding global development while appreciating China's influences and its perspectives. In the words of Mr Scott Morrison (Prime Minister of Australia), *"...it's obviously been the game-changing country of our era and our generation. That's obviously going to reset the world order..."*. The capacity for such substantial global dominance and influence has fostered significant interest from varying perspectives. Academics, policymakers, and industry practitioners alike are continuously becoming more inquisitive of the implications of China's emergence. This emergence alone is enough to give pause and warrant reflection. Any state that is able to exert such influence on the global economic and political structure will ultimately instigate a rebalancing of the geopolitical order.

But China cannot be viewed through the prism of just any state, and this is the underlying motif for this text. In fact, former US secretary of State Henry Kissinger once exclaimed that *"China's political culture has deep roots and is suffused with its own distinctive philosophical concepts of life, of hierarchy and of authority—a Confucian China with modern characteristics"*. This statement poignantly addresses the crux of objective assessments about China's political economy. The consequences, processes, and way of doing business in China are largely a foray into unchartered territory. It is on this basis that the relevance of academic pursuits about contemporary Chinese practices is even more objectively substantiated as they help to develop new knowledge. This reinforces the importance of ensuring objectivity and that matters about China are addressed with nuance and balance, to afford the reader the best opportunity of being comprehensively informed and not reinforcing any pre-existing biases they may have.

We, the authors of this manuscript, are thus aspired to attend to this need by offering new insights into China's trade policymaking, that is, international air transport policymaking, in a way not expounded hitherto. Taking a holistic and longitudinal approach, we have examined how China's international air transport policy goals are set and adjusted so as to ensure that the policy objectives are in alignment with the country's overall strategic diplomatic and economic development visions and how the goals have been evolving. Selecting a few country-pair markets for an in-depth analysis, we further investigated how the international air transport policy goals have been achieved and executed through bilateral air service agreements negotiations with its respective counterparts. Factors at different levels and from various sources have been identified and explored to justify what kind of influences these factors have had and how these factors have affected the policy direction and outcome. The narrative and analysis are based on the empirical evidence collected from primary data through dozens of interviews and secondary data from various sources such as files and records.

We have endeavoured to explicate Chinese international air transport policy with such detail and rigour that removes some of the mystique which has surrounded it. There is also a unique advantage to be gained from assessing air transport as a trade in service. On the one hand, policymaking in air transport is uniquely situated between a nation state's right to formulate domestic policy, and the broader dictates and influences it is subjected to within an international policy domain. In this context, it is likely that the global motivations and rationales that exist for liberalisation will prominently arise foremost in the minds of readers. On the other hand, however, it is important to consider whether this nexus has any peculiarities in the context of the Chinese approach.

This book is the culmination of endless hours of deliberations, writing, rewriting, editing, and overhauling. The result is a detailed recount of Chinese policymaking which has been dissected in such a way that the academic world has not hitherto been apprised of and which we are proud to share. Because of this, there were many causes for introspection and comprehensive deliberations. Scholarly ideas and practical evidence were mentally pitted against each other as we rationalised how the greatest justice could be served in assessing the realities in China. As with any pursuit, there were vacillations in this development which were premised in questions about how best the argument should be developed and fashioned. These hesitancies at those points emerged because there has been, and still remains, a constant commitment on our part to being true to our ultimate ambition; that is, a commitment to giving our readers the best opportunity to be exposed to and informed in a way that ensures a productive and fruitful pursuit while engrossed amongst these pages.

Nevertheless, we would like to acknowledge two notable caveats. Firstly, because of the project-based nature of the data collection, insights are only gleaned up to the year 2010. Secondly, data in relation to the China–Japan and China–South Korea air transport relationships has not been accessible due to various reasons. At the very least though, we are very proud to have contributed new insights which advance knowledge not only about China but also about the model applied to international air transport as a trade in service. While we hope that readers will meet with the same degree of elucidation, we commend to you this measured assessment for your own adjudication.

Before the final leaf is turned on this preface, there is one last point of note which will assist in manoeuvring through this discourse. It is easy to generate biases that the very nature of China's preceding closed structure would lead to heavy-handed diplomacy. This presumably would result in greater influence being wielded on international air transport by China's senior leaders. Indeed, national development took precedent and air transport was developed in response to these national dictates. The operative notion therein though is that it was responsive. In other nations, international air transport policy has been used as a proxy to create the balance necessary between domestic and foreign concerns. In some cases, such agreements become a means to some alternative diplomatic end. This would help to explain why liberalisation has been viewed with greater fluidity in western contexts. Conversely, in China, because air transport had largely been negated by senior political leaders as a source of political leverage during the period studied, bureaucrats apparently had greater scope in ensuring aviation policy reflected and enabled the national objectives. This gave rise to greater policy-based cohesion and overall consistency. In this context, the following remarks by Xi Jinping (President, Peoples Republic of China) are thus aptly fitting:

"Strength does not come from the muscle of the arms, but from the unison of the heart."[1]

On a more commemorative note, we would like to take this opportunity to similarly vest in this quote our immense sense of gratitude to all those who have participated in bringing this text to fruition. From the willingness of those who shared their insights and experiences in the very beginning, to the publishers who expressed profound resolution and commitment to seeing this brought to light, words of thanks pale in conveying the true sentiment of the authors. Now releasing this synthesis into the reader's domain, we hope that your takeaways as a reader entail the same objective and rewarding experience that we have had.

Associate Professor Dr Chrystal Zhang and
Mr Kareem Yarde (PhD Candidate)
Aerospace Engineering & Aviation
School of Engineering
College of Science, Engineering and Health
RMIT University
Victoria, Australia

Note

1 Ancient Chinese Proverb Remarks delivered at the Fourth Summit of the Conference on Interaction and Confidence Building Measures in Asia, 21 May 2014.

ACKNOWLEDGEMENTS

Nothing of merit is achieved in life without the support and contribution of others. Setting out to fill an intellectual void is no exception. Throughout this process, we have been blessed by the encouragement and assistance offered by many amazing and inspirational peers and industry professionals. We remain forever grateful and cannot thank them enough for their support and engagements. Special words of gratitude are due to Dr George Williams for his insightful guidance, the interviewees and industry professionals for sharing their insights without reservation, and our publishing team at Taylor & Francis. Specific mention is due to Guy Loft, Senior Editor, and Matthew Ranscombe, Editorial Assistant, for their patience and their belief in this project. Their support has been instrumental in helping us to bring this to fruition. Last but not least, our unwavering gratitude goes to our respective families, friends, and loved ones for their continuous support and contributions.

ABBREVIATIONS

AA	American Airlines (American)
ACC	Air China Cargo
ACI	Airport Council International
ACP	USA–China Aviation Cooperation Program
ADS	Approved Destination Status—appears in USA–China
AMM	ASEAN Ministerial Meeting
ANA	All Nippon Airways
AOC	Air Operator Certificate
APEC	Asia Pacific Economic Cooperation
ASA(s)	Air Services Agreement(s)
ASEAN	The Association of Southeast Asian Nations
ATAG	Air Transport Action Group
ATMB	Air Traffic Management Bureau of CAAC
BA	British Airways
BATA	British Air Transport Association
BCIA	Beijing Capital International Airport (Beijing Airport)
BRI	Belt and Road Initiative
BMI	British Midlands Airways
BMI	Business Monitor International
BSB	Beijing Statistics Bureau
CA	Air China
CAA	UK Civil Aviation Authority
CAAC	Civil Aviation Administration of China
CAB	Civil Aviation Board
CAFTA	China-ASEAN Free Trade Area
CATA	China's Air Transport Association
CCA	China Cargo Airlines (China Cargo)

CCP	Chinese Communist Party
CEPA	Mainland and Hong Kong Closer Economic Partnership Arrangement
CNSB	China National Statistics Bureau
CNTA	China's National Tourism Administration
CO	Continental Airlines
COSCO	China Ocean Shipping Group Company
CRS	Computer Reservation Systems
CZ	China Southern Airlines
DfT	Department for Transport of the UK
DL	Delta Airlines (Delta)
DoC	Department of Commerce of the USA
DoF	Department of Treasury of the USA
DoS	Department of State of the USA
DoT	Department of Transportation of the USA
EC	European Commission
EC	European Community
ECJ	European Court of Justice
EEA	European Economic Area
EEB/TRA	Transportation Affairs division in the Bureau of Economic, Energy and Business Affairs
ENDT	the Executive Management Development Training
EU	European Union
FAA	Federal Aviation Administration
FAM	Fragmented Authoritarianism Model
FedEx	FedEx Corporation
FFP	Frequent Flyers Programme
FTA	Free Trade Agreement
GATT	General Agreement on Tariffs and Trade
GBIA	Guangzhou New Baiyuan International Airport (Guangzhou Airport)
GDP	Gross Domestic Product
GE	General Electric Company
HSR	High-Speed Rail
HKAA	Hong Kong Airport Authority
HU	China Hainan Airlines Company Limited (Hainan Airlines)
IASED	International Aviation, Safety and Environment Division in the UK DfT
IATA	International Air Transport Association
ICAO	International Civil Aviation Organisation
IMD	International Institute for Management Development
IMF	International Monetary Fund
JAL	Japan Airlines
JAS	Japan Airlines System

JCCT	the Joint Commission on Commerce and Trade
JEC	the Joint Economic Commission
KAL	Korean Air
KLM	KLM Royal Airlines
LCC	Low Cost Carriers
MoFA	Ministry of Foreign Affairs of China
MFN	Most Favoured Nation
MoC	Ministry of Commerce of China
MoF	Ministry of Finance of China
MOFTEC	Ministry of Foreign Trade and Economic Cooperation of China
MoT	Ministry of Transportation of China
MOU	Memorandum of Understanding
MU	China Eastern Airlines (China Eastern)
NAFTA	Northeast Asia Free Trade Area
NDRC	National Development and Reform Commission
NPC	the National Planning Commission of China
NW	Northwest Airlines (Northwest)
OAA	Open Aviation Area
OAN	Office of Aviation Negotiations
O&D	Origin and Destination
OECD	the Organisation of Economic Co-operation and Development
OTP	Office of Transportation Policy
PRC	People's Republic of China
PRD	Pearl River Delta Region
RPK	Revenues Passenger Kilometres
RPM	Revenue Passenger Miles
RTK	Revenues Tonne Kilometres
RTM	Revenue Tonne Miles
SAA	Shanghai Airport Authority
SAR(s)	Special Administrative Regions
SASAC	State-Owned Assets Supervision and Administration Commission
SCO	Shanghai Cooperation Organisation
SDPC	State Development and Planning Commission
SED	Strategic Economic Dialogue
S&ED	the Strategic and Economic Dialogue
SEZ	Special Economic Zones
SOE	State-Owned Enterprise
SPIA	Shanghai Pudong International Airport (Shanghai Pudong)
Southwest	Southwest Airlines
TMM	Transport Minister Meeting
TWA	Trans World Airlines
UA	United Airlines (United)
UN	the United Nations

UNESCAP	United Nations Economic and Social Commission for Asia and the Pacific
UPS	United Parcel Services
USTDA	the United States Trade and Development Agency
VG	Virgin Atlantic Airways (Virgin)
WTO	World Tourism Organisation
WTO	World Trade Organisation
YRD	Yangtze River Delta Region

1

INTRODUCTION

International air transport as a commercial business constrained by politics

Air transport has emerged into an industry of consequential global importance, connecting the world and creating opportunities for trade and cross-cultural exchange. By 2017, more than 20,000 city pairs were established globally, which is an increase of 1,351 over the quantum in 2016 and a doubling of global connectivity since 1995, when there were less than 10,000 city-pair connections globally (IATA, 2018a). Increased connectivity and affordability have boosted trade in goods and services, and heightened foreign direct investment and other important economic flows. Although more than 99 percent of world trade by weight is by surface transport, more than one-third by value is transported by air, which is worth about $5.9 trillion, representing almost 7.5 percent of the world's GDP (IATA, 2018a).

Paradoxically, air transport was not invented to meet a clear, identifiable commercial demand. This is in contrast with the invention of other modes of transport, such as railways, which were developed to meet the needs of the British coal industry (Staniland, 2003). Instead, its genesis was to serve nations in the event of war to protect its territory and sovereignty and subsequently as a source of national prestige. Such a close link to national security and political power resulted in the industry being tightly controlled by sovereign governments (Karou, 2000). This relationship and its implications could be aptly summarised in the words of Lowenfeld as follows:

> International aviation is not just another problem in a changing economic system, though it is that, international civil aviation is a serious problem in international relations, affecting the way governments view one another,

the way individual citizens view their own and foreign countries, and in a variety of direct and indirect connections the security arrangements by which we live.

(Lowenfeld, 1975)

It was thus realised that aviation would have a greater influence on its foreign policy than any other non-political considerations, a warning issued in 1938 by the then Assistant Secretary of State, Mr Adolf Berle. Serving the political interest of a nation state was thus the overarching objective of any commercial aviation activity, with national carriers bearing a far more political significance. This close relationship between aviation and foreign policy is rooted in the premise that international air transport is viewed as an instrument akin to foreign policy based on reciprocity and privileges which sometimes have nothing to do with air transport (Sochor, 1991).

In 1944, air transport delegates from around the world convened in Chicago to formulate an international treaty that would establish rules to govern the industry. While they successfully addressed other areas, the delegates failed to agree on an arrangement that would allow airlines to pursue the economic benefits of operating in a liberated environment. In the absence of any permanent arrangements agreed by all participating countries, bilateral negotiations on cross-border airline operations and investment took precedent, which at least allowed the industry to fulfil its essential mission (Schenkman, 1955). The national government of the signatories, thus, have taken a key role in formulating their international air transport policies in line with the international regime.

They negotiate the so-called bilateral air services agreement (ASA), trading the specialised "traffic rights" of airspace access, with prioritised considerations given to the national sovereignty and security rather than the commercial demand for air travel. These traffic rights have gradually evolved into the nine freedoms of air traffic rights (Table 1.1) and are traded by sovereign governments in a resolutely bilateral fashion, with each side being committed to a kind of aero politics of restriction and artful compromise, classic zero-sum diplomacy, in defence of the home carrier's market share (Havel, 1997). Government bartering, not the entrepreneurial acumen of airline managements, has been the sole instrument driving new market development in this most technologically precocious industry over the decades (Havel, 1997).

These legally binding ASAs reached between governments tend to typically include commercial constraints with tight controls in the following areas:

- the number, size, and destination points of flights that an airline can operate between two countries;
- the type of direct connecting services or codeshare arrangements that an airline can offer;
- an airline's freedom to set their own fares, often requiring fares to be approved by one or both of the contracting nations; and
- majority of the ownership and effective control of an airline must reside with nationals of the relevant country (UK CAA, 2007).

TABLE 1.1 Nine freedoms of the air

Traffic rights	Definition
First freedom right	The right or privilege, in respect of scheduled international air services, granted by one state to another state or states to fly across its territory without landing
Second freedom right	The right or privilege, in respect of scheduled international air services, granted by one state to another state or states to land in its territory for non-traffic purposes
Third freedom right	The right or privilege, in respect of scheduled international air services, granted by one state to another state to put down, in the territory of the first state, traffic coming from the home state of the carrier
Fourth freedom right	The right or privilege, in respect of scheduled international air services, granted by one state to another state to take on, in the territory of the first state, traffic destined for the home state of the carrier
Fifth freedom right	The right or privilege, in respect of scheduled international air services, granted by one state to another state to put down and to take on, in the territory of the first state, traffic coming from or destined for a third state
Sixth freedom right	The right or privilege, in respect of scheduled international air services, of transporting, via the home state of the carrier, traffic moving between two other states
Seventh freedom right	The right or privilege, in respect of scheduled international air services, granted by one state to another state, of transporting traffic between the territory of the granting state and any third state, with no requirement to include such operation any point in the territory of the recipient state, that is, the service need not connect to or be an extension of any service to/from the home state of the carrier
Eighth freedom right	The right or privilege, in respect of scheduled international air services, of transporting cabotage traffic between two points in the territory of the granting state on a service which originates or terminates in the home country of the foreign carrier or outside the territory of the granting state
Ninth traffic right	The right or privilege of transporting cabotage traffic of the granting state on a service performed entirely within the territory of the granting state

Source: ICAO Manual on the Regulation of International Air Transport (Doc. 9626, Part 4).

The political and military considerations of international air transport have affected the growth of the industry. World trade cannot exist without the facilitation of transport, nor can it serve the common goals of nations and peoples unless it flows freely to every part of the world with minimum undue interference (Schenkman, 1955). The boundaries that slowed its progress were those invisible political ones which nations had to maintain for their real or fancied security, or to protect their national economy or international prestige and position (Schenkman, 1955). The close relationship between politics and the industry substantiated the continuous evolution of the type of global conversation that began with the Chicago conference. With political involvement rife in comparison with the more liberalised environments that exist today, some authors of the time vigorously expressed their contentions. Schenkman exclaimed, "It is the tragedy of air commerce that the nations of the world have never been able to agree on the kind of political boundaries needed in air space" (Schenkman, 1955).

The restricted bilateral mechanism thus constrained the ability of airlines to operate on a fully commercial basis and to achieve its operational efficiency. Firstly, the operational restrictions precluded airlines from entering into a market where there was a demand. Secondly, ownership restrictions hampered the airlines of one country from receiving capital investment from another country (Smyth and Pearce, 2007). These constraints prevented the air transport industry from mushrooming into a truly global industry. Moreover, they denied consumers and businesses the opportunity to witness the potential benefits which could be obtained from a fully liberalised industry. It is this latter point that propelled justifications of liberalisation in its nascent stage in the United States of America (USA). Thus, the liberalisation of the domestic air transport industry was enshrined within a broader quest to spur economic growth in the USA during the post-war period. Liberalisation would move the industry in the direction of one less constrained by regulatory barriers and more towards the application of free market principles.

Calling for liberalisation of international air transport

Governments were enjoined by the global economic growth that obtained in the 1980s to remove the restrictions on international trade and further promote the free movement of people and goods. In the USA, having been appointed as the chair of the then Civil Aeronautics Board (CAB) by President Jimmy Carter, professor of economics, Alfred Kahn spearheaded the dismantling of the system of government control over the industry as guided by the principle of free market economics. After deregulating its domestic airline industry and witnessing the attending benefits, the USA championed the liberalisation of international air transport operations since the early 1990s by concluding the so-called open skies agreements. Such a radical policy change in the USA was mainly triggered by the belief in the idea that liberalisation enables businesses to optimise their returns on investment through competition in a free market. Strong airline traffic in the 1970s and 1980s had enabled the US carriers to develop extensive domestic networks and improved their productivity and efficiency. Expansion into international markets with more traffic rights would better facilitate US carriers achieving economies of scale.

Convinced that the benefits of expanding into global markets for the US carriers could only be achieved through lifting the restrictions on traffic rights, the US government regarded open skies as the most effective mechanism to help sweep away the majority of the traditional limits on traffic rights and allow full pricing freedom, with an aim of creating a liberalised operational environment in international markets. Unlike standard bilateral agreements, open skies agreements eliminate the traditional roles played by governments in determining routes, designating the number of the airlines that could operate in the market, controlling the frequency of services, and approving tariffs. Instead, under an open skies agreement, market forces determine such matters. Airlines of one party may operate as many flights as it wishes so long as there is a need in the market, restricted only by safety concerns and airport capacity issues (Meyer, 2002).

Proponents of open skies contend that such arrangements stimulate airline competition, which in turn leads to lower fares and better service to the travelling public (Meyer, 2002). They argue that open skies permit aviation markets to grow in accordance with demand, thereby allowing airlines to offer more convenient and affordable air service to both passenger and shippers. This market growth would in turn also create economic expansion for the other party to the agreement. The oldest open skies agreement post-dated the Chicago conference by almost half a century when the USA entered its first iteration of the contemporary policy with the Netherlands in 1992. Evidence continues to suggest that the proponents of such agreements are correct. In numerous instances, growth in open skies aviation markets has significantly outpaced growth in similar markets governed by traditional, more restrictive bilateral treaties. More notably, service levels in open skies markets have been found to have increased while fares have fallen (Meyer, 2002).

However, open skies agreements have not been unopposed. Governments of nations that subsidised national flag carriers have feared the consequences of unrestrained competition against more efficient foreign carriers (Meyer, 2002). Other opponents include airlines themselves, which may benefit from their positions as incumbent carriers in markets being threatened due to the new entrants. Additionally, some airline industry observers have warned of the risk of "destructive competition" among airlines that may accompany a move towards a deregulated open skies environment (Meyer, 2002). This is well reflected in the examination by Button (1996), applying the concept of the core and the extent to which liberalisation can result in a collapse creating an empty core phenomenon. Notwithstanding, the divergent views have not stopped open skies being propagated across the world, particularly considering the resultant evidentiary growth.

When formulating policies to promote liberalisation, governments tend to have different considerations due to their specific social and economic systems, with an attempt to achieve their objectives with minimum opposition from various stakeholders. Different policies and approaches have been adopted in the exercise, resulting in different stages of the liberalisation process. Countries like Canada, Australia, New Zealand, and Singapore are among those first that have moved in concluding open skies agreements with the USA, thus becoming a fully liberalised market encouraging competition. The European Union (EU)

TABLE 1.2 Summary of key points in traditional ASAs, open skies, and OAA.

Type of agreement	Capacity and frequencies	Setting fares	6th to 9th traffic rights	Foreign ownership and controls allowed	Cabotage
Traditional ASAs	X	X	X	X	X
Open skies	Y	Y	Y	X	X
OAA	Y	Y	Y	Y	Y

Source: Adapted from CAA (2007).
(Remark: X represents the traffic rights that are restricted, while Y refers to the traffic rights that are allowed).

has liberalised its air transport market within its member states to its full extent without any restrictions on investment and operations, which goes hand in hand with the integration of its common market. Australia and New Zealand have created an open aviation area (OAA) which has removed all the restrictions on bilateral air transport operations. Developing countries in Asia, South America, and Africa, on the other hand, have been more conservative in accepting the open skies concept, taking a progressive and gradual approach towards the liberalisation of their international air transport markets. Table 1.2 summarises the differences between traditional ASAs, open skies, and OAA.

In an increasingly interconnected era where national economies are now more globally intertwined than ever, what is certain is that the future of international air transport does not reside in traditionally restrictive policies. As a driver of international trade, the dissolution of barriers in air transport fostered appreciable growth in other industries and sectors and thus further substantiated efforts of global initiatives like the World Trade Organization (WTO). The WTO emerged as a means of reducing regulatory barriers in international trade between nations by establishing globally agreed rules and benchmark standards. With services being responsible for two-thirds of the global economy, air transport service is as equally consequential as trade in any other service.

Trade in services and international air transport

Trade in services

Generally, trade is usually regarded as being beneficial because it allows a nation to specialise in the goods that it can produce relatively efficiently. This principle of "comparative advantage" is the essence of trade theory and is the foundation of free trade (Stiglitz and Charlton, 2005). Free trade is preferred because liberalisation of restrictions leads to a rise in the level of welfare and improves the average efficiency in a country, allowing resources to be redeployed from low-productivity protected sectors into high-productivity export sectors, and leading to a more integrated world economic system with globalisation being the main feature. Statistics show that from 1997 to 2007, merchandise trade grew at an average rate of 5.9 percent per annum. The WTO reports that since the global financial crisis in 2008, world merchandise trade has grown in tandem with GDP at a rate of 26 percent (WTO, 2019).

While businesses and consumers were enjoying the benefits, critical voices against free trade were becoming louder. Khor (2001), for example, asserted that colonial rules, as well as the imposition of new economic systems, have changed the social and economic structures, causing developing countries to grow more and more dependent upon global trading. It created an environment where countries were being forced to export more goods, mainly natural resources, and thus developed inextricable dependencies on the world economic system, consequently, losing their indigenous skills, capacity for self-reliance, confidence, and the very resource base upon which their survival depends (Khor, 2001). Morris (2001) argued that free trade is unable to promote or sustain the social relationships that create a vibrant community, as its key premise relies on a narrow definition of efficiency. He further pointed out that by taking precedence over the autonomy, sovereignty, and, indeed, the culture of local communities, the transport of capital, materials, goods, and people did little to contribute to the sustainable development of society (Morris, 2001).

Assisted by the rapid expansion of knowledge-based services such as accounting, advertising, marketing, and distribution, the efficient provision of which has become essential for corporate competitiveness, the separation of such activities from the production process has led to a vigorous development of trade in services. Communication technologies have facilitated the liberalisation efforts in capital and service markets, resulting in the shift of focus of international trade from manufacturing goods to services (OECD, 1997). In the USA, the manufacturing of goods dwindled to only 14 percent of its GDP in the early 1980s (Stiglitz and Charlton, 2005). As a response to this trend, "trade in services" was introduced in the Uruguay Round of discussions in 1986 by the General Agreement on Tariffs and Trade (GATT), later called the World Trade Organisation (WTO), which has since become the most dynamic element in world trade. Statistics from Organisation for Economic Co-operation and Development (OECD) showed that between 1980 and 1993, trade in services grew at 7.7 percent per year compared with 4.9 percent for merchandise trade, accounting for more than 50 percent of OECD countries' Foreign Direct Investment (FDI) outflows (OECD, 1997). Trade in services is particularly prevalent as an economic development tool employed by developed countries. This underscores the prospects for developing countries to improve their economic prospects by increasing their trade in services. Statistics from United Nations Conference on Trade and Development (UNCTAD) showed that in 2017, export of services trade grew at 13 percent for transition economies and 7 percent on average for developed and developing economies, which is in comparison to 11 percent growth for developing economies in merchandise trade relative to an 8 percent increase in trade for developed economies. Transition economies benefited most substantially from trade in goods, recording an increase of 24 percent on average (UNCTAD, 2017).

International air transport as a trade in service

Despite the different views and controversy associated with free trade, internationalisation and globalisation became two of the most pronounced industrial trends during the late 20th and early 21st centuries (Button, 2004). Air transport is closely linked to global trade growth and has contributed significantly to the ongoing economic integration on a global scale. Although the growth trend of world trade had been falling since 2003 and fell sharply in 2008/2009 as a result of the worst economic recession since the 1930s, air transport is responsible for 35 percent of international trade measured by their value though taking less than 1 percent by volume (IATA, 2018b). Air freight is also commensurately tied to the vagaries of global trade. This close relationship can be seen in Figure 1.1 which juxtaposes year on year growth of air transport freight with global merchandising trade volumes.

While facilitating the growth of international trade, air transport emerged as an important business in its own right. It created global connectivity by dissipating limitations once caused by geographical boundaries, thus making available formerly inaccessible utilities, bridging the geographical gaps between goods and services and consumers (Benson et al., 1994). Air transport contributed to the increase of human satisfaction by providing affordable travel for everyone in society, thus broadening their leisure and cultural experiences (Air Transport Action Group, 2005).

International air transport has since been regarded as a significant part of international trade in services being able to generate revenues to contribute to national foreign reserves. According to the International Monetary Fund's (IMF)

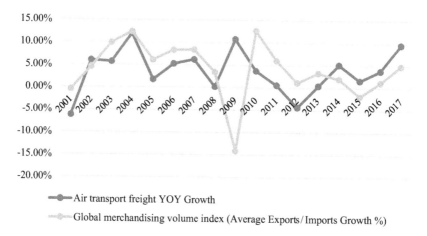

FIGURE 1.1 Relationship between global air freight tonne kilometres flown and world trade.

Source: Compiled by the authors from www.wto.org and www.unctad.org.

Balance of Payments Statistics Yearbook, trade in passenger air transport services is defined as passenger services provided by air transport and performed by residents of one economy for those of another. Therefore, a residency approach to trade in aviation services is generally consistent with IMF's balance of payments accounting. As Weisman (1990) explains, in principle, a country exports aviation services when an airline (a resident) from that country sells its service to a passenger (resident) in another country.

There is, however, no coherent dataset on trade in passenger air transport services for the world. A report by WTO, UNCTAD, and ITC Global estimated that in 2015, sea transport experienced a decline in exports by 12 percent compared to a more moderate decline of 7 percent in air transport. By 2017, the resurgence in exports by transport mode had air transport experiencing a year-on-year growth of 10 percent relative to the 5 percent growth in sea transport (WTO, 2018). Figure 1.2 shows the value of world imports in passenger transport services of all transportation modes, including air, shipping, railways, and others, between 2015 and 2017.

Air services are generally considered to occupy a large part of transportation, and one may assume that the trend of passenger transport services, as shown in Figure 1.2, will broadly reflect the trend of passenger air transport services. The value of world imports in passenger transport services, disclosed in the IMF's Yearbook, increased dramatically between 1988 and 2002, growing consistently with the exception of 2001, at about 7 percent per annum. Table 1.3 offers information on export and import values of passenger services in 2006 and 2013 for selected countries. In 2006, the USA, Germany, the UK, the Netherlands, and Spain were among the largest exporters. The USA had the largest export

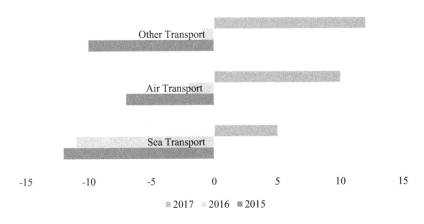

FIGURE 1.2 Transport services by mode 2015–2017.
Source: Compiled by the authors from WTO (2018).

TABLE 1.3 Passenger Services Trade 2006 and 2013

	Export / credit (Millions USD)		Import / debit (Millions USD)		Growth percentage	
	2006	*2013*	*2006*	*2013*	*Imports*	*Exports*
USA	21,640	41,640	22,640	32,030	41%	92%
UK	9,010	9,170	14,350	14,740	3%	2%
Germany	12,650	13,950	11,560	13,950	21%	10%
Japan	3,020	1,770	10,780	10,380	−4%	−41%
Canada	2,281	N/A	5,452	N/A		
Australia	2,870	2,272	4,756	6,622	39%	−21%
Italy	3,390	2,350	4,280	6,560	53%	−31%
Spain	6,24	7,224	3,651	6,278	72%	16%
Belgium	1,314	1,192	2,317	2,346	1%	−9%
Korea	2,720	5,015	2,138	2,330	9%	84%
Switzerland	2,044	3,457	1,947	3,028	56%	69%
Brazil	261	310	1,737	4,258	145%	19%
Thailand	3,221	3,960	1,575	1,735	10%	23%
Argentina	555	718	939	2,580	175%	29%
Malaysia	1,853	N/A	828	N/A		
Portugal	2,022	3,940	802	786	−2%	95%
Finland	1,135	1,702	675	1,065	58%	50%
Norway	498	7	552	511	−7%	−99%
Poland	883	1,069	430	303	−30%	21%
Egypt	542	883	372	419	13%	63%
Netherlands	6,147	7,073	366	287	−22%	15%
Chile	678	963	334	568	70%	42%
Peru	205	916	249	513	106%	347%

Source: IMF (2014).

value. Large importers included the USA, Japan, the UK, Germany, and Canada. From 2006 to 2013, exports and imports increased steadily in some countries, while other countries showed a different trend. Countries like Japan and Norway experienced declines in both import and export values in 2013 relative to their baseline year in 2006. Countries that demonstrated a large increase in exports were Peru, Portugal, Korea, and the USA, while those with a large increase in imports were Brazil, Argentina, and Peru. It is noteworthy that the most significant growth occurred in the emerging economies. European countries had initially enjoyed a large expansion of trade in passenger air transport services following the economic integration and the formation of a liberal single airline market in EU from the mid-1990s (Endo, 2007).

Recognising its importance to economic growth, national governments have taken a fresh view towards international air transport and have incorporated air transport policies into their overall economic policy considerations. The 1990s saw more countries taking actions to liberalise their international air transport markets considering the potential economic prospects (Doganis, 2002), while its political and military implications began to wane. This has generally obtained in a multitude of countries globally, particularly in the case of leading global industrial nations with few exceptions. Considering the trade potential to be derived from the air transport industry, assessing air transport policy in the context of trade theory can offer credible insights not hitherto explored.

In addition to viewing air transport in this way, there is also substantial value to be gained from exploring the evolution of the policy stance on international air transport in the context of one nation in particular, namely China. The 21st century is an epoch characterised by sustained threats to the global dominance of the USA by China. It is a dominance that has emerged on many fronts including in geopolitical, economic, and military realms. In the words of Joseph Nye, "not since Rome has one nation loomed so large above the others" (Nye, 2003). Consequently, this dominance created the context necessary to underpin the leading role the USA played in reshaping global policy positions in air transport. Admittedly, there remain much debate and contention about the potential transition of hegemonic power between China and the USA as a consequence of its rising global status and involvement in regional and global affairs (Zhang, 2004). Undeniably though, the American hegemony which has shaped much of the contemporary global society and economic policy, and in this case air transport policy, has been no more significantly challenged than now. For the foreseeable future, whether China emerges to replace the USA as the true global superpower, or whether it merely continues to pose a strong viable rival, it is unlikely that China's involvement in international air transport will be reflective of its involvement in the past. In effect, the China of today is not the China of yesterday. To this end, it is fitting that an analysis be performed which is focused on explicating the potential implications China's new global role can have for international air transport policy.

China's international air transport policymaking

China launched its economic reform with the aim of transforming its centrally planned economy into a socialist market economy. The policy change created the circumstances necessary to was to relieve the poverty in urban areas and to achieve modernisation for the nation which proved critical for its air transport growth. The burgeoning economic prospects directly implicated China's air transport policy through the increased trade and people-to-people exchanges spurred by its wider integration into the global community. China has experienced remarkable economic growth with an average growth rate of more than 10 percent per annum. When growth in the world economy was constrained by the global financial crisis in 2008, China remained one of the few countries maintaining 8 percent growth rate and overtook Japan to become the second largest economy with a GDP of 30.07 trillion yuan (US$4.42 trillion). According to the World Competitiveness Yearbook published annually by the International Institute for Management Development (IMD), of the 57 economies analysed,

China's overall competitiveness kept rising, with its world rank jumping from 18th in 2006 to 15th in 2007, thanks to its strong economic performance as a result of its structural reform and infrastructure investment. A decade later, advances in China's reforms and economic growth continued to sustain its enhanced global competitiveness, bringing the country's ranking to 13th in 2018.

In tandem with its expanding economy, air traffic grew rapidly over the decades with a massive increase in RPK from 2.79 billion in 1978 to 286.56 billion and an increase in RTK from 0.097 billion in 1978 to 11.77 billion in 2008. In 2008, China's air transport industry had employed 286,000 people and carried 191.9 million passengers, generating 304.7 billion yuan (US$45 billion) in revenue, ranking second among the ICAO member states, just behind the USA. Ironically, none of the Chinese airlines were ranked in the top 20 world airlines in terms of traffic carried or financial performance. In 2017, it accounted for the largest share of global RTKs (12 percent) rising from the previous 11.77 billion in 2007 to 84.5 billion in 2017, while RPKs rose to 951 billion, an increase of over 665 percent over the 2008 baseline. IATA expects that China will overtake the USA as the world's largest air transport market within the next decade.

Historically, air transport in China has been controlled and managed by the air force. This is because air transport is a part of the national defence system with military significance. International air transport agreements have been negotiated bilaterally and concluded only with those that recognised the status of the People's Republic of China (PRC) and established official diplomatic ties. Consequently, international air transport had been limited to only a few "friendly countries." The Civil Aviation Administration of China (CAAC), the government agency charged with the responsibilities of overseeing the air transport industry in China, reports to the Ministry of Foreign Affairs (MFA) and the State Council (SC) for policy advice on bilateral air transport agreement negotiations with its foreign counterparts. International air service is regarded as a political mission to strengthen bilateral relations. Hence, a reformation of the industry started only in 1987, with the corporate operations being separated from the government functions by creating six carriers based on CAAC's regional administrative operations. Numerous policies have been initiated and implemented afterwards, eliminating the restrictions on the economic aspects of the operations to encourage competition.

Although magnificent structural changes have taken place as a consequence of the political reform, China is still on its way towards full liberalisation. Believing that air transport is to a great extent an integral part of the nation's defence system, China holds that international air transport policy is an element of its foreign policy, which should be deployed to support the nation's overall strategic and military objectives. The turning point came in 2001 when China was accepted to the WTO. The government has since made more stringent efforts in liberalising its international air transport market. As China's approach towards international air transport evolves, it continues to make progressive steps towards liberalisation in stages. It can be argued that its destined position towards being the largest global air transport market discounts the extent to which the initial restrictiveness and protection of the domestic carriers could be maintained.

The underlying structural changes to China's airline industry were a result of the central government's political initiative which was fuelled in part by the transformation of national State-owned enterprises (SOEs) and the financial malaise of the carriers due to their noncommercial focus (Zhang 1998). This view of domestic conditions taking priorities in the decision-making of the government to intensify the industry reform was echoed in studies conducted by Zhang and Chen (2003), who argued that the domestic situation was the primary consideration when formulating liberalised policies. These considerations included low traffic demand from domestic passengers who were discouraged by the restrictive international travel policy and a lack of human capital and management expertise of the industry. State-owned large carriers, as a result of years of growth and industry consolidation, emerged to become monopolistic and oligopolistic firms at the expense of real market competition and were able to leverage their market power in influencing the government's policymaking process. Consequently, requests for market entry and increase of capacity from foreign governments tended to be rejected given the domestic circumstances. The multi-modal transport system and the increasing prevalence of high-speed rail has been expanding domestic connectivity while at the same time threatening the growth of domestic air transport options in these markets. In light of the domestic alternatives, Chinese airlines will face increasing pressure to redeploy capacity to international markets, magnifying interests from domiciled carriers in China's international air transport policy developments. On the other hand, the economic reforms in China and the trade and demand potential draws the interest of foreign carriers keen to gain market share. While there had been concern about the readiness of the domestic operators to compete effectively with their foreign competitors, generally, iterative progress towards liberalisation has been portrayed as China pursues its ambition of becoming an 'Aviation Power'. International Air Transport policy had been cautious and conservative prior to the early 2000s. During the periods of the economic reforms and external impetus by its global integration, the policy became increasingly liberal despite some inconsistency before 2010.

Studies on China's international air transport policymaking

There has never been a lack of interest in studying China's air transport market, with a wide array of literature documenting the industry's regulatory reform and its impact on the industry; however, there are several prevalent gaps within this body of literature.

Firstly, more attention has been given to the economic impact of the regulatory reform on the industry, in particular its domestic market. Scholars such as Zhang (1998), Shaw (2009), Wang and Jin (2007), Zhang and Zhang (2016, 2017), Zhang and Lu (2013), Law et al. (2018), and Wang et al. (2019) have shed light on the changing landscape of the sector and its contribution to the country's economic growth. However, the aspect of its international air transport market, especially its international air transport policymaking considerations, driving forces, and factors that affect its policy direction and policy objectives, is rarely examined.

Secondly, a couple of scholars analysed the country's aviation policymaking, but with a focus on domestic market. Dougan (2002) analysed the policymaking of the country's aviation industry from a political economic perspective but only focused on domestic issues, with little attention to its international air transport policy. Williams (2009) endeavoured to understand the driving forces for the changes of China's air transport industry from both domestic and international sources, but with limited analysis directed to its international markets. Undeniably, these collective efforts have contributed significantly to the understanding of China's air transport industry, its development and growth, and its prospects and challenges.

Thirdly, when taking international air transport as an element of international trade, it is regrettable to note that there have been little studies on the subject matter in China. Over the decades, sinologists such as Kenneth Lieberthal and Michael Okdenberg (1988), David Shambaugh (1994), Samuel Kim (1994), Quansheng Zhao (1996), Thomas Robinson (1994), and Yufan Hao (2006) have conducted substantial research on China, mainly focusing on its political, economic, and security aspects, and in particular, its bilateral relations with the major world powers such as the USA, Russia, Japan, and the EU. They have studied China from an international relation's perspective, examining and analysing how the country's foreign policy is made, the country's security considerations, and the implications with respect to its policy towards its bilateral partners and the rest of the world.

Since the 21st century, research on China's international foreign policy, including foreign trade policy, has been focused primarily on the country's interactions with its major trading partners such as the USA and Japan. Several other works have examined China's policymaking process by focusing on the international environment as well as its domestic politics with an attempt to assess the impact of these factors on the country's policy outcomes. For example, Feng (2006) examined China's entry into the WTO and argued that it was a state-led, leadership-driven, and top-down process, with the top leaders playing a decisive role in the course. Zeng and Mertha (2007) together with the contributors in their collection examined the impact of domestic sources on the country's foreign trade policymaking and argued that lobbying patterns in China are becoming increasingly similar to those observed in advanced industrialised states, with ranked agencies, ministries, sectoral interests, and even transnational actors becoming increasingly capable of influencing both the outcome of China's trade agreement as well as the terms of such agreements. Another strand of literature has been devoted to analysing the implications of China's accession to the WTO, assessing the impact of the incumbent social issues such as income disparity, social welfare, and industrial development on its WTO membership. What is missing from this strand is the evolution of its trade policymaking in the shadow of WTO membership (Zeng and Mertha, 2007). Their analysis has shed light on China's international trade policymaking process as well as its international behaviour (Zeng and Mertha, 2007). Their contributions are unprecedented in facilitating mutual understanding between China and the rest of the world.

Knowledge gap

The lack of study on China's international air transport sector and its policy-making, in particular from a political–economic perspective, can be attributed to the following. Firstly, the subject matter is interdisciplinary, requiring the examination of essentially domestic arguments about industry policies, as well as issues concerning international trade, and, indeed, international politics (Staniland, 2003). Secondly, there exist some constraints in terms of having access to empirical data research, due to various reasons, although there is a burgeoning source of memoirs and memorials, as well as interviews with those who know Chinese elite politics and key policymakers, which are becoming more readily accessible. Last but not least, there remains an absence of a comprehensive analytical framework to be applicable to such an analysis due to its unique positioning, i.e. the subject matter is caught between the domestic economy and foreign policy[1] (Cohen, 2000).

Research aim and objectives

The need to supplement incumbent scholarship has inspired us to attend the knowledge gap through a China-tinged lens.

We aim to explore the driving forces that have influenced China's international air transport policy direction and goals. We attempt to identify the factors from various sources that have affected the considerations of Chinese policymakers and examine how these factors have interacted with an effect on the policy goals. We also intend to investigate how the international air transport policy objectives have been achieved and executed in a country-pair market through bilateral ASAs negotiations and how the timing and circumstances have affected the policy outcome.

Specifically, the following objectives are established:

- To identify the considerations of the government in determining its international air transport policy direction and goals;
- To identify the driving forces and factors that shape China's international air transport policy direction and goals;
- To determine whether the identified factors are evolving;
- To establish how the country's international air transport policy goals have been achieved and executed on a country-pair market;
- To explore how and to what extent the bilateral ASA negotiations have affected the country's capacity to achieve its policy goals and outcomes; and
- To investigate whether there are any variations between the country-pair markets in terms of setting and achieving the country's international air transport policy goals.

Research design

A holistic and longitudinal approach

As China's integration into the world economy continues, international air transport will have a bigger role to play to contribute to Chinese "going global" strategy, and the policy therefore will, to a great extent, affect not only the air transport industry but also logistics and international trade at large. From the inception of its economic reforms in the 1970s, the Chinese aviation industry has been expanding in keeping with its widening economy and it is expected to grow to represent the largest air transport market within the next decade (Charlton, 2018).

International air transport policy is exemplified by a cross-border bilateral country-pair air transport services arrangement. The policy outcome is achieved through negotiations with the counterpart—from seeking information and formulating the agenda, negotiation strategy, and negotiation process, to concluding and implementing the agreements. All of these aspects are examined and analysed in the context of bilateral country-pair negotiations with China and other bilateral partners.

A holistic and longitudinal approach to its policymaking is also necessary since the evolution of air transport worldwide has involved both public policy and business strategy, and entailed decision-making that has occurred at international, national, and corporate levels (Staniland, 2003). China has a strategically important position in the contemporary geopolitical environment, which is largely premised in its relatively dilatory opening and integration into the global society. Consequently, it would arguably be conceptually deleterious to assess the policymaking in international air transport without relying upon an arduous and substantive chronicle. To do so could inadvertently create the circumstances where confirmation bias could evolve in relation to perceptions about liberal or restrictive policies, depending on the prevailing impetus at that particular time. In other words, if a singular defining moment in time is used as the focal point (such as the period of WTO entry), the ability to assess how the various influences have interacted with each other over time would be implicated. The discussion would then only surround the drivers and circumstances attending at that particular period. Naturally, the analysis must be delineated by some particular epoch. But the argument herein is that there is greater benefit to be had if the timespan chosen is representative enough that it allows for the consideration of multiple influences on the policy development.

China's economic growth and the overall transitioning of its society and economy have been perpetuated over time by the development of strategic five-year plans. Consequently, these strategic plans are critical to China's national policymaking and the overall positioning of any industry therein. During the 10th Five-Year Plan, a major milestone in liberalisation was reached. For the first time, there was an explicit declaration to liberalise the international air transport industry (Yang, 2005). Then in 2005, the 11th Five-Year Plan signalled another landmark when private investment into Chinese carriers was permitted (Williams, 2009). In the first decade of the 21st century, these two policy positions represent significant nascent evolutionary movements in Chinese aviation. This text therefore focuses on China's international air transport policies until 2010 which coincides with the end of the 11th Five-Year Plan.

Examining China's international air transport policymaking and its execution through bilateral country-pair ASA negotiations

International air transport policy is not unilateral, though developed internally. It is often affected by various bilateral and international circumstances. China's international air transport policy is reflected in its policy towards another country with which it negotiates an agreement. The different social, political, and economic systems across nations, especially the divergent market conditions, shape the policy with respect to each country-pair market. It is the array of these bilateral arrangements which is derived from the various negotiations that reflect a nation's overall policy and policy change with respect to international air transport. Thus, a comprehensive analysis of China's policy changes for different country-pair markets aids in identifying contextual implications on the policy. To this effect, analysis was conducted using case studies of select bilateral partners, which include the China–USA, China–UK, and China–Netherlands markets. Additionally, some discussion in the context of China's peripheral neighbours Japan and South Korea helps to present a useful juxtaposition.

Such an approach to China's international air transport policymaking on a country-pair market would allow researchers to identify the international factors such as the international regimes that have been governing international air transport and bilateral political and economic relations. Further, it better highlights the domestic considerations of decision makers, their preferences and choices, the impacts of interest groups and the media, and the relationships established with other bureaucracies and institutions. In addition, institutional changes and the key decision makers are included in this analysis to determine how political reform in the restructuring of institutions has affected the functions of the organisations and the extent to which the institutional context and individual decision makers are influencing the final policy outcome. To further understand the subject matter, factors at both the international and domestic levels are examined, analysed, and compared to help understand their changing nature and implications for the policymaking process.

Other considerations to focus the research on country-pair air transport markets included the following:

• It has been heavily regulated by international regime bilaterally as well as by national governments, and is treated differently from other industries;
• It is a truly commercial business, generating economic benefits to individual nations and the world as a whole, but carries significant political and national defence implications;

- It is part of international trade, facilitating globalisation, but itself is still undergoing the process towards full liberalisation; and most importantly
- It best reflects the country's evolving approach to liberalisation in the international air transport sector.

Selected country-pair markets as case studies

In an attempt to seek answers to the aforementioned questions, a case study approach is adopted. A case study is an in-depth investigation of a discrete entity (which may be a single setting, subject, collection, or event) on the assumption that it is possible to derive knowledge of the wider phenomenon from intensive investigation of a specific instance or case (Thornhill et al., 2009). It is an empirical inquiry that investigates a contemporary phenomenon in depth and within its real-life context, especially when the boundaries between phenomenon and context are not clearly evident (Yin, 2009). It is particularly suited to research questions which require a detailed understanding of a social or organisational process because of the rich data collected in context. It is the ability to deal with a full variety of evidence such as documents, artefacts, interviews, and observations, beyond what might be available in a conventional historical study, that is the unique strength, which aids to serve the purposes of this research which attempts to investigate a particular event to answer the questions of "how" and "why" (Yin, 2009).

Five country-pair markets in different geographical areas which have had different socioeconomic and political relations with China were selected for in-depth analysis. From North America, attention is paid to the China–USA relationship. This juxtaposition is substantiated not only by the geopolitical and economic positioning of the two nations but also by the leading role that the USA has played in driving the liberalisation of air transport at domestic and international levels. The European relationship is unique because of its intricately developed integration movement which affects bilateral agreements between EU and non-EU member states. As discussed in Chapter 5, the EU-based horizontal agreement requires a mandate from member states, and in this case, no such mandate had been received during the period analysed. Therefore, European representation is provided through the Netherlands and the UK. The Netherlands has long adopted a liberal stance to international air transport in support of its international and open economy. As one of the most liberal members of the EU, analysing relations in the case of the Netherlands would thus offer a substantive perspective. The China–UK air transport relationship offers one of the most long-standing and storied point of views to be assessed across the EU membership. There is also a complexity in the bilateral relations in the case of Hong Kong SAR that would be precluded in an analysis revolving around any other EU member state. Within Asia, attention is paid to Japan and South Korea as two other leading economies within the region. In contrast to the foregoing western nations, the Japanese and South Korean relationships are more constrained by issues of confidentiality. It should therefore be noted that information in these contexts is less explicit and the analysis is less comprehensive.

Data collection

To optimise the strengths of data collection methods available for a case study (Yin, 2009), both primary and secondary data were collected through qualitative techniques. The primary data were collected through more than 20 in-depth face-to-face interviews with government officials who have been involved in the bilateral ASA negotiations, diplomatic officials involved in facilitating bilateral communications, local government officials responsible for facilitating air transport planning and development, airlines executives participating in the negotiations, airport managers responsible for international route development, and academic peers whose expertise is in aviation and international relations research. The interviewees included senior government officials, including European Commission, airline and airports representatives, and officials from industry association. All interviewees have abundant personal experience in both/either formulating national international air transport policy goals and/or participating in numerous country-pair bilateral air service agreement negotiations, thus enabling them to share insight for this research.

The secondary data, such as documents and records in both English and Chinese, were collected from various sources in the public domain, which included but were not limited to government reports, journals, industry magazines, corporate brochures, annual reports; internal memos such as Minutes of Meetings and Memoranda of Understanding, negotiations reports and press releases, circulars and notices, proceedings and conference papers; and news clippings and mass media articles. The evidence gathered from the document review is enormously valuable as it provides a wide range of written records of the happenings, as well as extensive background information allowing for unmitigated factual accounts.

Analysis framework

A Micro-Macro Linkage Approach[2] has been applied to analyse the international trade policymaking. The analysis framework, developed by Zhao (1996), would enable the authors to identify the factors that influenced China's international air transport policymaking from international, regional, domestic, and institutional levels and further examine how these factors have evolved and interacted with each other, thus affecting the policymaking process and the policy outcome on a country-pair basis. In this way, this book brings academic value by supplementing the available literature on international air transport policy in China.

Book structure

The remainder of the book adopts the following structure. Chapter 2 describes the characteristics of China's air transport sector. It discusses the economic reform and air transport sector regulatory reforms that took place in China, which facilitated the growth of its air transport. With a burgeoning domestic market, attention is then directed towards China's liberalisation process from 2001. As liberalism in China continues to expand, the connectivity of Chinese airports is also explored.

Chapter 3 discusses the theoretical framework in which this analysis of the policymaking is rooted. It starts with the concept of international trade policymaking and explains its interdisciplinary nature, which is straddling the domains of both foreign and economic policies. It discusses a variety of different theories applied by both economists and political scientists who have attempted to analyse the policymaking process, and examines the pros and cons of each theory. It describes in more detail the Micro-Macro Linkage Approach developed in the 1990s by Zhao and argues that it is believed to be the most appropriate model to be applied to analyse China's international air transport policymaking.

Chapters 4–6 are the depictions of five case studies which include the China–USA, China–Netherlands, China–UK, China–South Korea, and China–Japan country-pair markets. Each case study discusses the various rounds of bilateral negotiations between China and the respective counterparts, exploring and analysing the considerations and expectations of the policymakers before the consultations, the strategies and tactics adopted by both sides, as well as the outcomes agreed following the consultations. They then compare the agreements reached out of each round of discussions and analyse the motives of both parties for coming to a compromise. In this way, the chapters are positioned to identify the various factors and assess how the identified factors have affected the policymakers who eventually make the decisions and come to an agreement with their counterparts.

Chapters 7–9 are the application of the micro-macro linkage model for an explicit analysis of the afore-mentioned country-pair case studies. It identifies the factors at international, regional, and bilateral levels; politically, economically, strategically, and operationally; and discusses how these factors influenced the country's national and industry policy directions and objectives. It then continues to identify the factors at national, societal, institutional, and individual levels and examines how these factors react to the factors at the international, regional, and bilateral levels; how they interact with each other and converge on the policymakers. It analyses critically each of the factors identified, explores how each of them has affected the mindset and considerations of policymakers in formulating the policies, strategies, and tactics before, during, and after the negotiation process and assesses its weight in influencing the policymaking process. It further establishes the evolving nature of the factors and determines how the time- and environment-constrained nature of the factors has influenced the policymaking process. By comparing and contrasting the identified factors, it showcases how international air transport policymaking in China is influenced by all factors at all levels that shape its transnational air transport market.

Chapter 10 concludes by summarising and critiquing the identified factors that have influenced China's ever-evolving international air transport policy direction and goals. While academic contributions are garnered from the application of the model and the case studies, recommendations to policymakers in terms of developing a more sound and systematic approach in formulating the international trade policy with respect to international air transport are also derived. In this regard, stakeholders should be more involved in the policymaking process to ensure that the policy outcome reflects the wider interest of the society, rather than a single industry. From an academic perspective, limitations of the model and a suggested revised framework are also identified.

Notes

1 Please refer to Chapter 3 for an explicit discussion of the theoretical model adopted for this research.
2 Please refer to Chapter 3 for an explicit discussion on this model.

References

Air Transport Action Group, (2005), The economic & social benefits of air transport, available at www.icao.int/atworkshop/ATAG_SocialBenefitsAirTransport.pdf

Benson D., Bugg R., and Whitehead G., (1994), *Transport and logistics*, Woodhead-Faulkner (Publishers) Limited, England.

Button K., (1996). Liberalising European aviation: is there an empty core problem? *Journal of Transport Economics and Policy*, 30(3), 275–291.

Button K., (2004), *Wings across Europe, towards an efficient air transport system*, Ashgate Publishing Ltd., Aldershot.

Charlton E., (2018). China is building 8 new airports every year, World Economic Forum, available at www.weforum.org/agenda/2018/08/these-five-charts-show-how-rapidly-china-s-aviation-industry-is-expanding/

Cohen S. D., (2000), *The making of United States international economic policy: principles, problems, and proposals for reform*, Praeger Publishers, Westport, CT.

Doganis R., (2002), *Flying off course: the economics of international airlines*, (3rd edition), Routledge, London.

Dougan M., (2002), *A political economy analysis of China's civil aviation industry*, Routledge, London.

Endo N., (2007), International trade in air transport services: penetration of foreign airlines into Japan under the bilateral aviation policies of the US and Japan, *Journal of Air Transport Management*, 13, 285–292.

Feng H., (2006), *The politics of China's accession to the World Trade Organization*, Routledge, London and New York.

Havel B. F., (1997), *In Search of open skies: law and policy for a new era in international aviation*, Kluwer Law International, aan den Rijn.

Hao Y., (2006), Introduction: influence of social factors: a case of China's American policy making, in: Hao Y., and Su L., (editors), China's foreign policy making, societal force and Chinese American policy, Ashgate Publishing Limited, Aldershot, the UK

IATA, (2008), IATA economic briefing, the impact of recession on air traffic volumes, December 2008, available at www.iata.org/whatwedo/Documents/economics/IATA_Economics_Briefing_Impact_of_Recession_Dec08.pdf

IATA, (2018a), IATA annual review, available at www.iata.org/publications/Documents/iata-annual-review-2018.pdf

IATA, (2018b). *IATA cargo strategy*, IATA, Geneva.

Karou D., (2000), *Lobbying the European Commission, the case of air transport*, Ashgate Publishing Limited, Aldershot.

Khor M., (2001). Global economy and the developing world, in: Goldsmith E., and Mander J., (editors), *The case against the global economy & for a turn towards localization* (pp. 146–155). Earthscan Publications Ltd, London.

Kim S. S., (1994), China's international organisational behaviour, in: Robinson T. W., and Shambaugh D. L., (editors), *Chinese Foreign policy: theory and practice*, (pp. 401–434), Oxford University Press, Oxford.

Law C. C., Zhang Y., and Zhang A. (2018). Regulatory changes in international air transport and their impact on tourism development in asia pacific. in: Fu, X., and Peoples J., (editors), *Airline Economics in Asia* (pp. 123–144). Emerald Publishing Limited, Bingley.

Lieberthal K., and Oksenberg M. (1988), *Policy making in China: leaders, structures, and processes*, Princeton University Press, Princeton, NJ and Oxford.

Lowenfeld A. F., (1975), A new takeoff for international air transport, *Foreign Affairs*, October 1975: 36, available at www.jstor.org/stable/pdfplus/2706476.pdf?acceptTC=true

Meyer G. S., (2002), US-China aviation relations: flight path toward open skies? *Cornell International Law Journal*, 35, 427.

Morris D., (2001), Free trade: the great destroyer, in: Goldsmith E., and Mander J., (editors), *The case against the global economy & for a turn towards localization.* (pp. 115–126). Earthscan Publications Ltd., London.

Nye Jr, J. S., (2003), *The paradox of American power: why the world's only superpower can't go it alone*, Oxford University Press, New York.

Organisation for Economic Co-operation and Development (OECD), (1997), *The future of international air transport policy, responding to global change*, L'avenir Du Transport Aerien International, Paris.

Robinson T., (1994), Chinese Foreign Policy from the 1940s to the 1990s, in: Robinson T. W., and Shambaugh D. L., (editors), *Chinese Foreign policy: theory and practice*, (pp. 555–602), Oxford University Press, Oxford.

Schenkman J., (1955), *International civil aviation organisation*, Librairie E. Droz, Geneve.

Shambaugh D., (1994), Patterns of interaction in Sin-American Relations, in: Robinson T. W., and Shambaugh D. L., (editors), *Chinese Foreign policy: theory and practice*, (pp.197–223), Oxford University Press, Oxford.

Shaw S., Lu F., Chen J., and Zhou C., 2009, China's airline consolidation and its effects on domestic airline networks and competition, *Journal of Transport Geography* 17, 293–305.

Smyth M., and Pearce B., (2007), IATA economics briefing No 7, Airline Liberalisation, April 2007.

Sochor E., (1991), *The politics of international aviation*, Macmillan Press Ltd., Basingstoke.

Staniland M., (2003), *Government birds, air transport and the state in Western Europe*, Rowman & Littlefield Publishers, Inc., Lanham, MD.

Stiglitz J., and Charlton A., (2005), *Fair trade for all, how trade can promote development*, Oxford University Press, Oxford.

Thornhill A., Saunders, M., and Lewis, P., (2009), *Research methods for business students*, Prentice Hall, London.

UK Civil Aviation Authority, Economic Regulation Group, (2007), Connecting the continents, long haul passenger operations from the UK, CAP771, (July 2007), available at www.caa.co.uk/publications

UNCTAD, (2017), Trade and development report 2017: Beyond austerity towards a global new deal, UN, Geneva.

Wang J., and Jin F., (2007), China's air passenger transport: an analysis of recent trends, *Euroasian Geography and Economics*, 48 (4): 469–480, Bellwether Publishing, Ltd.

Wang H., Yang., J., and Wang., H., (2019), The evolution of China's international aviation markets from a policy perspective on air passenger flows. *Sustainability*, 11, 3566.

Weisman E., (1990), *Trade in services and imperfect competition, application to international aviation, international studies in the service economy* (Vol. 2), Kluwer Academic Publishers, Dordrecht, Boston, MA and London.

Williams A., (2009), *Contemporary issues shaping China's civil aviation policy, balancing international with domestic priorities*, Ashgate Publishing Ltd, Farnham.

WTO, (2018), Highlights of World Trade 2017.

WTO, (2019), *World Trade statistical review 2019*, World Trade Organization, Geneva.

Yang Y., (2005), Remarks by Minister of the General Administration of Civil Aviation of China (CAAC) at the 2005 ICAO Symposium on Liberalization of Air Transport in Asia/Pacific, ICAO, available at www.icao.int/Meetings/LiberalizationSymposium/Documents/2005_Icao-Symposium-Liberalization-Shanghai/Papers/Yang_Speech_en.pdf

Yin Robert K., (2009), *Case study research and applications: design and methods*. Sage Publications, Thousand Oaks, CA.

Zeng K., and Mertha A., (2007), Introduction, in: Zeng K., (edited), *China's foreign trade policy, the new constituencies*, Routledge, Abingdon.

Zhang A., (1998), Industrial reform and air transport development in China, *Journal of Air Transport Management*, 4, 155–164.

Zhang A., and Chen H., (2003), Evolution of China's air transport development and policy towards international liberalisation, *Transportation Journal*, 42(3), 31–49.

Zhang B., (2004), American hegemony and China's US policy. *Asian Perspective*, 28(3), 87–113.

Zhang Y., and Zhang A. (2016). Determinants of air passenger flows in China and gravity model: deregulation, LCCs, and high-speed rail. *Journal of Transport Economics and Policy (JTEP)*, 50(3), 287–303.

Zhang Y., and Zhang A. (2017). Air transport development: a comparative analysis of China and India, in: Finger, M., and Button K., (editors), *Air transport liberalization: a critical assessment* (pp. 112–137). Edward Elgar Publishing.

Zhang, Y. and Z. Lu (2013), Low Cost Carriers in China and its Contribution to Passenger Traffic Flow, *Journal of China Tourism Research*, 9, 207–217.

Zhao Q., (1996), *Interpreting Chinese foreign policy, the micro-macro linkage approach*, Oxford University Press (China) Ltd., Hong Kong.

2

CHARACTERISTICS OF CHINA'S AIR TRANSPORT INDUSTRY

Yahua (Shane) Zhang, Chrystal Zhang, Kareem Yarde[1]

Overview of China's economic growth and air transport industry

Air travel and economic growth are inextricably linked. On the one hand, there is a cyclical nature to air transport since the demand for air travel depends heavily on economic conditions. On the other hand, it is a widely held view that as an input into many economic activities including tourism, trade, and investment; air transport has been an important component in achieving economic development and welfare enhancement (Zhang and Findlay, 2014). Air transport is particularly important to distant and remote regions where there is no close substitute for this transport mode due to distance. In some parts of the world, air transport is the only viable means of transportation for both goods and people due to geographic or climatic constraints (Pagliari, 2010). This co-relationship is best reflected by leisure travel, which heavily relies on the increase in disposable income, while at the same time, air transport can substantially contribute to a country's tourism by bringing in tourists and revenues, thereby increasing local residents' disposable income.

China has been the second largest aviation market in the world in terms of the volumes of passengers and air cargo movement in its domestic market since 2007. In 2018, the whole industry handled 611.7 million passengers and 7.4 million tonnes of air cargo, representing 10.9 and 4.6 percent increases from the previous year, respectively. China's airline market is a growing market underpinned by a huge population and rapid economic growth. IATA forecasts that China will overtake the United States of America (USA) as the largest air passenger market in the mid-2020s measured by traffic to, from, and within a country. It is believed that the growth in China and other Asian economies including India, Indonesia, and Thailand will shift the centre of gravity of the air transport industry from the west hemisphere to Asia in the next two decades (Figure 2.1).

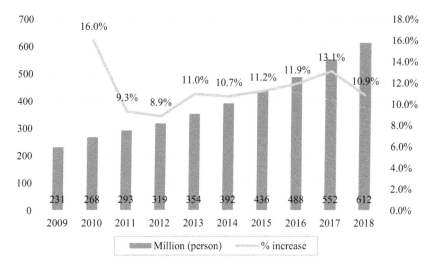

FIGURE 2.1 Number of passengers handled by China's airline industry, 2009–2018. *Source*: CAAC.

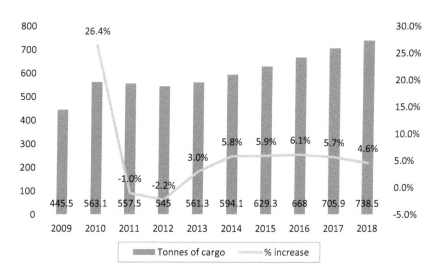

FIGURE 2.2 Tonnes of cargo handled by China's airline industry, 2009–2018. *Source*: CAAC.

Figure 2.2 depicts the passenger and cargo traffic carried by China's air transport sector from 2009 to 2018. The growth rates for both passenger and cargo markets were remarkable immediately after the economy recovered from the global financial crisis. In recent years, the growth rates for both markets were quite stable. The passenger market performed particularly well, recording a two-digit growth, even though the GDP growth rate was only around 7 percent in the past few years.

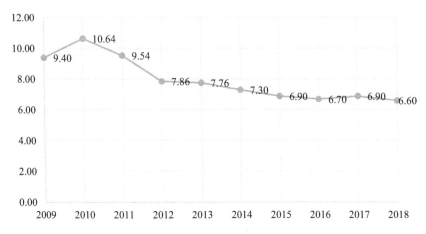

FIGURE 2.3 GDP growth rate % between 2009 and 2018.
Source: CAAC.

The country's GDP had grown from $216.5 billion in 1978 to $3.6 trillion in 2007 with an average growth rate of 9.8 percent per annum (National Bureau of Statistics, 2008), much higher than the world average of 3.0 percent for the same period of time. It is also higher than Japan whose growth rate was recorded at 9.2 percent during its economic take-off period and South Korea which reported 8.5 percent economic growth for the same period. The IMF estimates that China's GDP was ranked fourth in 2007[2] contributing to 6 percent of the world total, compared with its ranking of tenth in 1978. Accordingly, the country's average growth rate in air passenger traffic between 1985 and 2005 was recorded at 15.8 percent (Wang and Jin, 2007). Almost half a decade on, between 2009 and 2018, the highest GDP growth of 10.64 percent was recorded in 2010 following the resurgence after the global financial crisis and expansionary monetary policies adopted globally. Since then however, GDP growth has slowed and has consistently declined reaching a decade's low of 6.6 percent in 2018 (Figure 2.3).

Overview of China's multi-modal transportation system

China's remarkable economic growth has been accompanied by the extensive development of the country's transportation system (Zhou and Szyliowicz, 2006), in which the central government has invested heavily with an aim of promoting export-oriented economy and improving the quality of life. Such a growth is reflected in the increase of passenger volumes carried by all modes of transport except waterways, which has experienced a steady decline. Although road transport still accounts for more passengers, air transport has seen a massive increase of traffic. The China Statistical Yearbook details that in 1978, Chinese airlines carried only 2.3 million passengers, compared with 814.9 million passengers by rail and 1,492 million passengers by road. Approximately three decades later, in 2008, airline passengers increased to 192.5 million, compared with 1,462 million by rail and 26,821 million by road (National Bureau of Statistics of China, 2018). By 2017, air traffic had expanded at the largest annual rate to 551.6 million, representing a year-on-year growth of 15 percent across the period relative to 6 percent year-on-year growth of highway traffic growth during the same period. Table 2.1 shows the changes in passenger traffic between 1978 and 2017.

Corresponding with the passenger traffic, freight transport has experienced unprecedented growth too. According to the China National Statistics Bureau (CNSB), all transportation modes realised 19.7 trillion revenue tonne kilometres (RTK) in 2017. This represented an 8 percent increase in year-on-year growth compared with 983 billion RTK in 1978. Of all transportation modes, air cargo transport experienced double-digit growth rate, much higher than all others with the exception of highway freight (Table 2.2) (National Bureau of Statistics, 2008).

Accordingly, the market shares of the different transport modes have changed. In 1978, rail transport played the biggest role in moving traffic, taking more than 60 percent market share, with air transport accounting for only 1.6 percent, a tiny slice of the market. In 2007, rail transport had lost almost half of its market

TABLE 2.1 Passengers carried by different transportation modes in select years between 1978 and 2017 (in millions)

Transport mode (in millions)	1978	1990	1998	2008	2017
Highway	1,492.3	6,480.9	12,573.3	26,821.1	14,567.8
Rail	814.9	957.1	950.9	1,461.9	3,083.8
Water	230.4	272.3	205.5	203.3	283
Air transport	2.3	16.6	57.6	192.5	551.6

Source: stats.gov.cn.

TABLE 2.2 RTK and growth rate between 1978 and 2007

	RTK (in billion tonnes kilometre)		Change (%)
Year	1978	2017	
All modes	983	19,737.3	8.0
Rail	534.5	2,696.2	4.2
Highway	27.4	6,677.1	15.1
Water	378	9,891.1	8.7
Air	0.1	24.3	15.1
Pipeline	43	478.4	6.4

Source: www.stats.gov.cn.

share plummeting to only 33.42 percent, with air transport gaining a 12.93 percent with an average annual growth rate of 15.8 percent compared with 9 percent of other transport modes.

The development of high-speed rail (HSR) and its impact on air transport services

In the past decade, high-speed rail (HSR) has emerged as a significant transport mode in China, posing a serious threat to China's air transport sector because of its potential to reduce air travel demand (Zhu et al., 2019b). In 2018, the length of China's HSR track amounted to 27,684 km, representing the largest HSR network (64 percent of the total) in the world. According to the updated 'Medium-to-Long-Term Railway Network Plan' report covering the period 2016–2025 with an outlook to 2030, China's HSR network will by 2025 stretch to 38,000 km, including eight north–south and eight east–west trunk lines (Fu et al., 2015). By 2030, China's HSR network will reach 45,000 km in length, and most cities with a population of 500,000 or more will be connected by HSR. In fact, under China's recent expansion plan, by 2025 about 80 percent of its domestic airline routes are to be overlapped with HSR lines (Zhang et al., 2019).

The spread of HSR network has forced Chinese airlines to cut domestic airfares and reduce or cancel flights (Zhang et al., 2019). Zhang and Zhang (2016) showed that the presence of HSR services would significantly reduce the bilateral air passenger flows by 53 percent. In fact, in extreme cases air services could be suspended as a result of the launch of HSR service: 48 days after the opening of the HSR between Zhengzhou and Xi'an, all the flights between the two cities were cancelled; in the same year and for the same reason, airlines withdrew from the Wuhan–Nanjing and Wuhan–Nanchang routes; Wuhan Tianhe Airport recorded its first negative growth (−8.52 percent) in air passenger throughput during the Chinese New Year holiday in 2011 due to the impact of the opening of Guangzhou–Wuhan HSR. Quite consistent results are reported in the work by Chen (2017) who investigated the air-HSR competition on the Wuhan–Guangzhou and Beijing–Shanghai routes and found a significant drop in air traffic, flight frequency, and seat capacity as a result of the introduction of parallel HSR services. Specifically, the author reports a drop in domestic passengers by 28.2 percent, in flight number by 24.6 percent, and in seat capacity by 27.9 percent after the introduction of HSR services. The negative impacts on air service are the greatest on the routes with a distance between 500 and 800 km. When the Guangzhou–Wuhan HSR opened in 2009, there was a decline of air services between the two cities by 45 percent and the fall was 33.6 percent when the Beijing–Shanghai HSR was launched. Li et al. (2019) again confirmed the strong negative impact of HSR frequency on air-travel demand. Such a negative impact of HSR is stronger in China's central and western regions.

As a strong competitor to airlines, HSR is expected to put a downward pressure on airfares. Interestingly, however, mixed results have been produced regarding the impact of HSR on airfares. Ma et al. (2018) found that before 2014, HSR did place a negative pressure on airfare in China's airline market. However, from 2014, the negative impact gradually disappeared. The authors explained that two reasons are behind the changes. First, airlines were unlikely to charge higher prices immediately after the launch of HSR services, but they could develop strategies over time to respond to the HSR entry. Reducing frequency and capacity is one example. Seeking price-fixing is another possibility. Ma et al. (2019) found that the entry of HSR led to airfare convergence on the Beijing–Shanghai airline route, which might indicate the existence of tacit collusion among the operating carriers. Second, HSR could complement air services and bring in more passengers from nearby cities at both endpoints of the route. The director of the Civil Aviation Administration of China (CAAC) announced in 2017 that some flights from second- and third-tier cities to Beijing would be shifted to Tianjin and Shijiazhuang with these airports being linked to Beijing via HSR (Zhang et al., 2017). Ma et al. (2019) pointed out that while HSR poses a threat to air transport, it can also be used to mitigate congestion problems at mega-airports and help make full use of secondary airports' capacities through an air–rail cooperation agreement. It is expected that the deepening of such cooperation would help stabilise airfares.

In the face of a strong and irreversible competitor, HSR, that offers similar products, China's air transport sector needs to work out new strategies and develop new policies to sustain the growth of this industry. Obviously, encouraging price-fixing activities is no longer a choice as it is illegal to do so under the 2008 Anti-Monopoly Law, although explicit price collusion was a common practice in China's aviation market. For example, after the control over airfares was relaxed in 1997, the CAAC constantly sought to re-regulate airfares to avoid price wars and industry losses by putting a limit on the maximum discounts (Zhang and Zhang, 2017), including introducing a "revenue pooling" programme and supporting airline consolidations to restrict competition from the late 1990s to the early 2000s (Zhang and Round, 2008). With the introduction of the anti-trust law and the HSR services, Chinese airlines have a stronger incentive to lobby the CAAC not to open heavily travelled and lucrative markets to new carriers. This has been the case as mentioned earlier and can be justified at this stage given the fact that the airports of Beijing and Shanghai are over-congested. However, with the opening of the new Beijing Xiaxing Airport and the introduction of the third terminal in Shanghai, congestion is of a less concern, and there is no legitimate reason not to treat the state-owned and private carriers equally. As such any market access restrictions may not be able to stay long.

Zhang and Zhang (2016) pointed out that the challenges facing China's air transport sector will be greater in the near future after the rapidly expanding HSR network has connected most of the major cities. Therefore, Chinese airlines need to consider redeploying part of their capacities to international markets. This is actually the case for Spring Airlines that has shifted a significant part of its capacity on to the East Asia and Southeast Asia markets in the past few years. This implies that the Chinese government needs to consider embracing more liberal air services agreements/arraignments (ASAs) including actively pursuing 'open skies' deals. This will be discussed in the next section.

China's air transport policy direction, policy goal, and policy statement

Unlike European countries and the USA which have a clear vision with respect to air transport policy, Asian countries have divergent policies and objectives. Some advocate a free and liberalised market while others are not sure as to what they want to achieve, in particular, the less developed countries (Forsyth et al., 2006). In the case of China, the objective of its air transport policy has been changing over the past three decades, with varying enthusiasm for liberalising the market, as a response to both internal and external circumstances. The industry, thus, has gone through several uncertain periods of deregulation, which has boosted growth but is then followed by tight regulatory actions taken by the central government in attempts to control the overheated development (Williams, 2009).

Regardless, the Chinese government sees the air transport industry as one of the strategic industries, which plays a pivotal role in its endeavours to position itself as a powerful entity with its own characteristics. In the past few decades, its air transport policy direction and goals have experienced a few adjustments to align with the country's strategic objectives.

No explicit policy goal in the 1950s–1970s

Before the 1980s, few aviation-related laws and regulations had existed, nor any explicit aviation policy statements. Air transport had been a component of the national defence force, serving the country's national defence interest. Commercial air transport service was a luxurious and privileged treatment and was only available for those who were eligible and authorised for such exclusivity.

Aiming to become an "aviation-big" country during the 1980s and early 2000s

The economic reform started in 1978 marked a new era for the air transport industry in China. To align with the country's overall economic growth objective, for example, opening to the outside world and attracting international investment to stimulate a strong economic growth, the Chinese government for the first time spelt out its aviation policy goal. It asserted that "air transport should provide fundamental air transport infrastructure and satisfy air travel demand with its ultimate goal being to facilitate the country's economic development." By corporatizing the whole industry through separating it from the military and setting up financially independent carriers, the government encouraged the industry to expand to serve the needs for economic growth. In 2004, China was ranked the third largest aviation country among 188 International Civil Aviation Organization (ICAO) member states, thanks to its rapid growth in just more than two decades.

Aspiring to transform from "Aviation Big" to "Aviation Power" after 2002

The Chinese government has become more vocal and outspoken since 2002 advocating its aviation policy direction and goals, which have been illustrated in a series of key policy statements including the National Five-Year Plan and Civil Aviation Industry Five-Year Strategic Plan. Several landmark policy statements in the early 2000s included "China Civil Aviation Reform and Restructuring Plan"(2002), "The State Council's Notice with respect to Civil Aviation Regulatory Reform Plan", and the CAAC's "2003 China Civil Aviation Development Policy Statement". "For the first two decades in this new 21st century, our goal is to achieve the transformation from an aviation-big to an aviation-power country", as the plan declared.

However, the Chinese government was still in the process of exploring the strategies to be taken to become an "aviation power". Liberalisation had not been an acceptable word to use, the whole concept being a very sensitive issue and one not widely accepted across China. As a consequence, domestically, the government was responding to the rapid growth in a passive way, with laws, rules, regulations, and procedures being issued for rectification of industry misconduct. Internationally, each bilateral agreement was dealt with on a case-by-case basis, thereby giving rise to flexibility but also uncertainty (Forsyth et al., 2006). The lack of detailed strategic plans and guidelines at the government level resulted in ambiguity, which sent mixed and confusing messages to the industry, which in turn had to react passively to any challenges from both within and outside China. The situation has been improved in the next decade, which saw the issuance of several five-year plans, including the 13th Five-Year National Economic Development Plan, the 13th Five-Year National Strategic Emerging Industries Development Plan (issued by the State Council in November 2016, and The 13th Five-Year Civil Aviation Development Plan of China issued by the CAAC in February 2017. These five-year plans articulated an ambitious vision for the air transport industry and outlines the blueprint for the sector to follow.

Regulatory reform of the industry

Relaxing economic controls on domestic market

Relaxing economic controls on the domestic air transport market went hand in hand with the country's overall economic reform. By introducing a series of measures to lift the restrictions on investment, route licensing, pricing, and airport charges, the preliminary aim of the government was to ensure that air transport was able to provide convenient and efficient links across the nation, thus facilitating economic growth. The first measure was taken in 1984, and the CAAC allowed more airlines such as Xiamen Airlines and Shanghai Airlines to be established in 1985 and 1986, respectively, with the funding from both local government and other state-owned enterprises in the region. This was followed by the establishment of seven trunk airlines on the basis of the operations of the CAAC's regional administration in seven cities, that is, Air China (CA) in Beijing, China Eastern (MU) in Shanghai, China Southern (CZ) in Guangzhou, China Northern in Shenyang, China Northwest in Xi'an, China Southwest in Chengdu, and Xinjiang Airlines Urumqi. By 1995, there were 42 airlines operating in the country, some only having a fleet of less than five aircraft.

The provision of air transport services significantly contributed to the local economic growth but at the same time caused severe safety concerns, with nine aircraft crashes recorded between 1992 and 1993. The inadequate safety management system, a serious shortage of experienced pilots, a lack of management experience, and ineffective government regulations were to be blamed for the tragedies, which resulted in the CAAC's tight control in approving any new airlines. Measures were also taken to encourage consolidation with smaller airlines being taken over by trunk carriers, leaving the total number of air operators at less than 30.

The tight control on market entry remained unchanged until 2004, when the CAAC responded to the call of the national government that private investment was welcomed in the air transport industry, including airlines, airports construction, and ground handling services. It was also a response to the central government's commitment to the WTO, which accepted China's membership in November 2001. Tens of regulations, orders, programmes, and proposals were issued between 2004 and 2008 to relax the economic restrictions in terms of market entry and price control in the domestic market (Table 2.3).

These measures prompted a second wave of rapid growth in the number of airlines similar to that of the late 1980s and early 1990s. New carriers, such as Spring Airlines, Juneyao Airlines, and Okay Airlines with private capital, were approved in 2005 to operate in the domestic market. Joint-ventured cargo operators such as Great Wall International[3] and Jade Cargo International[4] were permitted in 2004 to serve all cargo international routes. Before 2004, there were 27 carriers operating in China, while by the end of 2007, there were 41 in operation. The increase of airlines brought about a dramatic increase of capacity and services. In 2003, there were a total of 961 domestic routes connecting 148 airports and 194 international routes connecting 72 cities in 32 countries. By the end of 2008, the 41 carriers operated on 1,235 domestic routes connecting 156 domestic cities and 297 international routes connecting 104 international cities in 43 countries (CAAC, 2008). Passenger traffic grew from 87.59 million in 2003 to 185.76 million in 2007, and RPK rose from 126.3 billion in 2003 to 279.2 billion in 2007 (CAAC, 2008).

TABLE 2.3 List of key regulations with respect to domestic operations issued by CAAC between 2004 and 2008

Date of entry into force	Regulation title	Key points	Reference number
6 February 2004	Proposed Programme to Speed up the Development of Air Cargo	• Allowed establishment of all cargo airlines • Allowed the development of cargo hubs • Relaxed cargo route licensing • Allowed flexible cargo rates within an approved range	CAAC YUN (2004) 28
20 April 2004	Pricing reform programme with respect to domestic air transport	• Allowed market-determined pricing for regional and feeder routes, and flexible pricing within an agreed range on trunk routes	CAAC 2004–18
1 January 2005	Regulation on Air Operator's Certificate	• Allowed establishment of commercial civil aviation enterprises	CAAC Order No. 138 (CCAR–201)
15 August 2005	Regulation on Domestic Investment in the Civil Aviation Industry (Provisional)	• Allowed private investment in the air transport industry, including the setting up of commercial airlines and the construction of airports and investment in air traffic control systems.	CAAC Order No. 148 (CCAR–209)
20 August 2005	Regulation on Merger and Restructuring of Aviation-related Business	• Allowed restructuring and mergers of aviation-related businesses • Allowed IPO and equity restructuring • Allowed contract management	CAAC Order No.149 (CCAR–229)
20 March 2006	Regulation on Domestic Route Licensing	• Allowed airlines to operate on most of the domestic, regional, and feeder routes by filing at the CAAC and their regional branches, unless otherwise specified	CAAC Order No.160, (CCAR–289TR–R1)
28 December 2007	Airport Charges Reform Programme	• Airport charges based on categories of airport operations. • Aeronautical charges decided by the government • Non-aeronautical charges determined by the market within an agreed range	CAAC 2007.158
12 March 2008	Procedures in Managing Domestic Route Licensing of Air Transport	• 45 days prior to the actual operation to lodge the application for the route licence • 30 days for processing after receiving the application • The licence will be revoked should no flights be operated within 60 days of the issuance of the licence, or the actual operation was less than 50% of the original application.	N/A

Source: Compiled by the authors from the information on www.caac.gov.cn.

The mushroomed increase of airlines put a strong challenge to the state-owned carriers and raised security and safety concerns. The CAAC then decided to suspend approval of new domestic entrants until 2010. Following the crash of an aircraft of a local airline in 2010 (Yichun aircraft crash), the government extended the suspension policy until 2013. Subsequently, another wave of private airlines emerged in 2013 and 2014.

It is argued that 2014 was a turning point for the Chinese low-cost carrier (LCC) sector because of the release of the 'Guiding Opinions on Promoting Low Cost Aviation Industry Development' by the CAAC. For the first time, the aviation authorities acknowledged the significant role played by LCCs in the nation's economy. From 2013 to 2014, there was another wave of private airlines established in China. Some of the existing carriers rebranded themselves as LCCs during this period, including China United Airlines (see Table 2.4). At the end of 2018, there were 45 state-owned airlines and 15 private airlines. Among the 60 carriers, 9 of them were all cargo carriers and 8 were publicly listed. Ten carriers had foreign equities. The chequered development journey of China's private airlines is shown in Figure 2.4.

Some of the private airlines established in the first wave around 2005 quickly failed due to the lack of experienced pilots and skilled personnel, and the high costs and taxes associated with aircraft procurement, jet fuel, and airport charges (Zhang and Lu, 2013). However, the most serious obstacles that stifled the growth of China's private and LCCs are the hostile aviation policy. In most cases, the state-owned airlines could exercise a significant influence on the CAAC's decision. Any aviation reforms and new aviation policy would put the interests of the state-owned airlines first (Zhang and Zhang, 2016, 2017). For example, for a long

TABLE 2.4 Profile of LCCs in China as of December 2017

Airline name	Airline code	Year declared as LCC	Fleet size	Base	Ownership
Spring Airlines	9C	2005	81(A320, B737)	Shanghai Hongqiao and Pudong, Shijiazhuang, Shenyang, Shenzhen	Private
West Air	PN	2013	30(A319, A320)	Chongqing, Zhengzhou	Private
China United Airlines	KN	2014	31(B737, B738)	Beijing	State-owned
Jiuyuan Airlines	AQ	2014	14(B738)	Guangzhou	Private
Lucky Air	8L	2016	45(B737, B73G)	Kunming, Lijiang, Chengdu	Private

Source: Websites of relevant airlines.

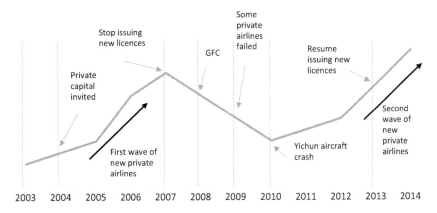

FIGURE 2.4 Chequered development journey of China's private airlines, 2003–2014. *Source*: Adapted from Xia (2014). (GFC for "global financial crisis.")

time, airport slot allocation in China has been a closed-door deal that favoured the state-owned airlines. When the Shanghai-based Spring and Juneyao were granted the right to fly between Shanghai and Beijing, they were only given a departure time from Shanghai late night and from Beijing early morning, almost the last two flights from Shanghai to Beijing and the earliest ones from Beijing to Shanghai. The big three operated about 50 flights every day while these two private carriers only operated one flight each. Being unable to attract many passengers, Spring suspended its service on this route for some time and did not return until recently.

Yu et al. (2019) compared the operating efficiency performance between Chinese and Indian carriers. They found that China's three state-owned airlines performed poorly in both the capacity generation and service stages, particularly the latter. In contrast, the private LCC, Spring was one of the most efficient carriers between 2005 and 2015. They confirm that the LCC model and private ownership are significantly associated with better airline efficiency performance. Interestingly, the state-owned Air India is much more efficient than its Chinese counterparts, probably indicating that state-owned airlines operating in an environment dominated by private competition and LCCs tend to become stronger in efficiency. Therefore, there is a need to formulate supportive policies towards LCCs and private carriers in China. In 2018, China further eased investment access to aviation industry, allowing private capital to account for more than 50 percent of their equity as long as the government remains to be the largest single shareholder. This move will likely improve the efficiency of the state-owned carriers.

Liberalising the international air transport market

The 1944 Convention on International Civil Aviation, also known as the Chicago Convention, established the ICAO as the governing body for the aviation industry worldwide. A regulatory framework, including three elements, was formed to deal with the economic issues involving international air services: bilateral ASAs for the control of market access; inter-airline commercial or pooling agreements; and IATA for controlling tariffs (Doganis, 2001). A typical ASA specifies the right to fly across borders and such rights are restricted to airlines designated by signatories. These airlines need to be owned and controlled by residents of the country making the designation. As a result, airlines from third parties are discriminated against: they either cannot fly the routes between the countries involved in the bilateral agreement or have only restricted access. Some bilateral agreements also restrict the capacity and frequency of the services which the designated airlines provide.

A cautious and conservative approach until early 2000s

Liberalising the international air transport market has been a long, gradual, and ongoing process for China. Without recognising adequately the instrumental role that international air links could play in facilitating the country's export-oriented economic growth, the CAAC's primary concern in the early days of the reform was to ensure that the newly formed carriers were able to optimise the advantages gained from their respective geographical locations so that there was minimal overlapping operation on an individual route to minimise head-on competitions.

To this effect, the CAAC developed economic and administrative measures to control potential competition. For example, market entry was controlled by the CAAC by stipulating clearly the scope of business activities in each of the carriers' air operator certificate (AOC), specifying what kind of business they were allowed to be engaged in and what type of routes they were entitled to operate on. For example, Air China, the only national flag carrier, inherited all the international traffic rights from the CAAC, was entitled to engage both passenger and cargo operations and to operate on both domestic and international routes but mainly focusing on long-haul destinations in the USA and Europe. China Eastern was given the right to operate on both domestic and international markets but with a small number of long-haul international routes to the USA in addition to a few East Asian countries. China Southern was allowed to operate in both domestic and international markets but mainly focusing on routes connecting peripheral countries in Southeast Asia. The other three carriers were only permitted to operate domestically within mainland China, without access to Hong Kong or Macau Special Administrative Regions (SARs). With such an arrangement, the CAAC was able to make sure that there was little overlap in the route network, thus the interests of each carrier being adequately protected. It would also help to avoid any dispute in international route allocation, as the AOC would not allow non-eligible carriers any undesired international expansion.

Under such circumstances, international operations developed only slowly, with bilateral ASAs being concluded as an element of the national strategy for the establishment and maintenance of diplomatic ties. For example, there were only 25 international routes for Shanghai in 1998 and 18 for Guangzhou, the capital city of Guangdong which was regarded as the powerhouse of the country's export industry. The flag carrier was the core consideration for the government as to whether there would be a need to relax any restrictions in terms of market entry, capacity, and pricing. Reciprocity was the philosophy governing the bilateral negotiations, with arrangements made with foreign counterparts being mainly protectionist and similar in all bilateral markets. The agreement with the USA in 1980, allowing double designation of carriers from each side to operate on two routes, respectively, was very unique and special at that time, given the fact that there existed no second carrier in the market for the Chinese government to designate. It was not until 1992 when China was able to designate China Eastern as the second carrier in the market, which launched the US-bound services with its newly delivered long-haul aircraft.

The shift to a moderately liberalised policy began in the early 1990s, when a spectacular growth in air transport was experienced in the market and more aircraft delivered. Some carriers such as China Southern, China Southwest, and China Northern were not happy about their restricted business scopes and started to lobby the government for international expansion. To satisfy their needs, double designation and flexible capacity arrangements were successfully negotiated, in particular with several Asian countries such as South Korea, Japan, and Singapore. However, favourable treatment was only extended to those CAAC-controlled airlines that demanded overseas expansion, despite the fact that a total of more than 40 airlines of different sizes and capabilities had been operating in the market.

A progressive but inconsistent liberal approach after 2001

The radical shift towards a more liberalised international air transport policy came after November 2001 when China was admitted to the WTO as a member. The fundamental consideration of the CAAC was to deliver its series of important commitments to opening its market and to liberalising its regime to meet the WTO rules (Freeman and Lam, 2002). The WTO membership exposed the country fully to the international business environment, where its industry had to compete with those from developed, developing, and underdeveloped countries. The government recognised the urgent need for a regulatory reform in its agenda to eliminate barriers to promote free trade. Another impetus was the intention of creating strong SOEs in the airline industry to reduce competition, optimise the resources, and improve productivity and performance so as to get the industry prepared for international competition.

For the first time in October 2003, the CAAC stated explicitly to the international aviation community that its air transport policy was to liberalise the market with a "proactive, progressive, orderly and safeguarded" approach (ICAO, 2013). The principle was written into the Annual Strategic Development Plan for 2004, with the CAAC advocating its support to carriers that wished to expand into international and regional markets. An assessment mechanism for international market entry and exit was established with an aim to effectively allocate, manage and monitor the implementation of the international air traffic rights. In the same year, Hainan Province was allowed unlimited 3rd, 4th, and 5th freedom traffic operations, and Xiamen and Nanjing 5th freedom cargo operations by both Chinese and foreign carriers, though only on a pilot basis, with the primary objective being to facilitate the local economic development while at the same time to earn time for the CAAC to justify the political, economic, and social impact of the liberalisation policy on the industry and the society as a whole.

To further reinforce its stance that liberalising the country's international air transport market was its firm and long-term strategic objective, the CAAC issued policy guidelines in 2007 stating that international air services would be encouraged to connect Mid-Western and Northeastern regions to the rest of the world. Efforts would be devoted to supporting those carriers which were able to optimise the bilateral air traffic rights available to strengthen aviation relations with African, Central Asian, and Latin American countries (CAAC, 2007). Further consolidation of the airline industry was to be encouraged to enable the formation of larger and more competitive enterprises. The three traffic hubs in Beijing, Shanghai, and Guangzhou were to be developed and strengthened.

Although air transport has been instrumental in facilitating China's regional development and export-oriented economic growth, the lack of clear, coherent, and well-articulated policy objectives in terms of international air transport before 2003 left the country's airline industry unable to develop a competitive edge compared with their counterparts in their neighbouring countries, let alone carriers in the USA and Europe. The case-by-case approach adopted in bilateral ASA negotiations not only enabled the government to be flexible in adjusting its strategies and tactics (Forsyth et al., 2006) to optimise outcomes in light of the prevailing situations but also resulted in a circumstance where, to some extent, the government has had to accept the outcome of negotiations as a result of competing considerations. Reciprocity has, above all, been the key principle in the bilateral negotiations, dictating the outcome of the discussions.

Of the 110 ASAs China had signed by the end of 2007, only 28 contained more relaxed traffic rights allowing multiple designation and unrestricted capacity (CAAC, 2008). East and Southeast Asian countries, including Japan, South Korea, and Singapore, were among those that enjoyed more relaxed air transport terms of trade with China. African countries were also granted no restrictions in terms of designation and capacity for both passenger and cargo operations, including 5th, 6th, and 7th freedom traffic rights (CAAC, 2008). The ASA with the USA, following the successful conclusion of the 2004 and 2007 protocols, allowed the phasing in of multiple designation, freedom of commercial arrangements, and elimination of restrictions on cargo operations. The two protocols served as milestones, taking China's international air transport liberalisation

TABLE 2.5 Number of flights (yearly) of the big three, 2006–2018

	2006	2008	2010	2012	2014	2016	2018
Air China	1,704	1,944	1,936	2,432	4,284	5,280	6,032
China Eastern	888	960	1,200	1,560	3,052	4,780	5,072
China Southern	656	628	492	672	832	2,672	2,632

Source: IATA AirportIS.

process into a new era which is markedly reflected in the growth of China's big three carriers. Table 2.5 reports the number of flights of China's big three in the China–USA market from 2006 to 2018. It can be seen that Air China is the largest carrier in the market, but China Eastern has followed closely in recent years.

Charting a more liberalised international air transport sector since 2010

The "proactive, progressive, orderly and safeguarded" approach towards liberalising the international air transport market has resulted in incremental liberalised arrangements between China and its counterparts. The pace has been accelerated in the past decade since 2010. By 2016, a total of 118 ASAs had been concluded, but the numerous amendments thereto were made in attempts to remove economic constraints so as to allow the industry to operate in a more liberalised environment. In particular, significant progress has been made between China and its peripheral individual neighbour state as well as regional blocks such as ASEAN. The China-ASEAN deal inked in 2010 aimed to establish an unlimited air service arrangement (passenger and cargo) between China and ASEAN members. Since then, the number of flights between ASEAN and China has increased rapidly. Traditionally, air transport services between ASEAN and China were offered by the flag carriers, and between gateway cities. The open skies agreement allowed both flag and non-flag carriers to increase flight frequency and offer flights to many second- and third-tier cities (Law et al., 2018). As a result, air connectivity between ASEAN and China has increased substantially. The number of flights operated by China's big three between China, Thailand, and Singapore is shown in Table 2.6. It can be seen that all the three airlines experienced substantial increase in capacity in the China–Thailand market. Air China recorded a decrease in the number of flights between China and Singapore and the other two saw moderate increases. Usually, Singapore is a business destination, while Thailand is a tourist destination. It seems that tourist destinations benefit most from the China–ASEAN open skies.

Oum and Lee (2002) discussed the possibility of creating open skies in Northeast Asia and identified many obstacles. Reluctance of the state-owned carriers was one of them. However, more than a decade on, most of the obstacles have changed or disappeared. The benefits of the open skies and the single aviation

TABLE 2.6 Number of flights operated by the big three between China, Thailand, and Singapore, 2007–2017

Airline	From China to	2007	2009	2011	2013	2015	2017
China Southern	Singapore	1,176	904	1,848	2,156	1,936	2,632
	Thailand	2,184	1,504	2,840	5,980	12,236	11,680
Air China	Singapore	2,568	2,352	2,672	2,016	1,968	1,968
	Thailand	1,244	940	884	1,824	3,776	5,760
China Eastern	Singapore	2,400	2,184	3,504	3,688	3,096	3,560
	Thailand	2,084	1,812	4,204	7,128	10,196	9,512

Source: IATA AirportIS.

market examples have been observed and accepted by many governments and consumers, which makes the conclusion of an open skies zone in Northeast Asia more possible than in the early 2000s. Liu and Oum (2018) note that the rapid growth of China's big three has conferred them with the opportunity to play a leadership role in the world air transport sector.

There have been regular meetings among the aviation authorities of the three Northeast Asian economies. Open skies arrangements have also been implemented between Chinese Shandong province and Korea since 2006. China and Japan reached an open skies deal in 2012 but this deal excludes flights to/from Beijing, Shanghai, Tokyo Haneda, and Tokyo Narita. In 2019, China and Korea signed an expanded bilateral air services MOU to add 14 weekly flights between Beijing and Seoul to support Beijing Daxing International Airport that is to be open in later 2019. Korea is keen to pursue an open sky deal in this region, given its relatively small domestic market and its close cultural and economic links with China and Japan. The signing of an open skies agreement between Japan and Korea in 2007 has lifted restrictions on frequency, capacity, and destinations, with the exception of the congested Tokyo airports, covering both cargo and passenger services. The number of flights to and from China operated by the major carriers in Northeast Asia is shown in Table 2.7.

As can be seen from the table, Air China and China Eastern recorded a decrease in the number of flights between China and Korea. However, one may argue that it is possible that these airlines may have used larger aircraft and thus reduced the frequency. A closer look reveals that the types of aircraft used have been quite consistent. Interestingly, Korea's two major airlines had substantial increases in the number of flights in the China–Korea market. In the China–Japan market, China's big three reported a steady increase while Japan Airlines showed a decreasing trend in the number of flights. For any liberal arrangements, attending benefits can be unequally yoked. However, the quantum of capacity directed towards the Northeast Asian market implies that the benefit of open skies or a single aviation market is likely significant.

TABLE 2.7 Number of flights to and from China to Japan and South Korea, 2007–2017

Airline	Between China and Japan and South Korea	2007	2009	2011	2013	2015	2017
Air China	Japan	11,248	11,324	11,544	11,596	14,104	14,952
	South Korea	8,080	6,240	6,784	7,000	8,784	6,472
China	Japan	11,872	12,888	13,756	13,604	20,440	23,112
Eastern	South Korea	17,404	8,612	10,256	12,152	15,812	13,912
China	Japan	8,432	7,016	7,040	4,952	8,104	8,888
Southern	South Korea	9,888	8,276	9,480	11,692	14,800	12,192
Korean Air	South Korea	14,504	14,352	15,652	18,064	19,752	19,884
Asiana	South Korea	15,532	15,636	18,164	20,164	21,416	20,272
Japan Airlines	Japan	14,400	13,924	8,456	8,680	8,848	9,408
ANA	Japan	12,488	12,016	12,280	13,664	14,216	16,688

Source: IATA AirportIS.

Oum and Lee (2002) argue that even in a bilateral negotiation, it is difficult to achieve air transport liberalisation unless the flag carriers of both countries are equally strong and competitive. Table 2.7 shows that the major carriers in the three countries do not differ much in terms of presence in the Northeast Asia markets. Therefore, it might be the right time now for the three countries to seriously consider creating an open skies zone in this region.

In 2009, the CAAC introduced the one-route, one-Chinese-carrier policy on the long-haul international routes to prevent cut-throat completion between Chinese carriers, particularly during the economic downturn. The long-haul routes refer to those with a distance of more than 4,500 km and are mainly the routes from China to the USA and Europe. In 2018, the policy was revised in the advent of the opening of the second international airport in Beijing and at a time when more and more Chinese could afford overseas travel and Chinese carriers had increasing interests in opening new international routes. The new Measures on "International Traffic Rights Resource Allocation and Use" aims to establish a sound, open, fair, and just management system for international traffic rights resource allocation and use.

The new measures divide international routes into two classes. Class 1 air routes are those from China to countries with open skies or partial open skies agreements. These countries include ASEAN countries, Australia, Chile, Maldives, Georgia, the USA, the UK, New Zealand, Norway, Denmark, and Sweden. For Class 1 routes, there are no restrictions on the number of designated carriers, flight schedules, frequency, and transport capacity. Class 2 international air routes are the markets with no liberal bilateral arrangement which are divided into long-haul Class 2 international air routes and non-long-haul Class 2 international air routes. Routes to the USA, Europe (excluding Russia), Oceania, and Africa are long-haul Class 2 international routes. A competition mechanism will apply for those (including new entrants) that want to fly on these routes. That is, a point system will be used to decide who will be the winners among the applicants. Other routes are called non-long-haul Class 2 international routes and there is no limit on the number of designations. This policy has drawn wide attention and sparked much discussion as it represents a new milestone of China's international air transport. The implementation of this policy will undoubtedly increase competition and drive down prices on some long-haul international routes.

China has also relaxed its arrangements with Oceanic countries such as Australia through several round of negotiations. China is Australia's largest trading partner in terms of both imports and exports, while Australia is China's sixth largest trading partner. China is Australia's second largest inbound tourist market after New Zealand, and the largest total expenditure market. Air transport between Australia and China has experienced phenomenal growth in the last ten years, with more direct flights launched between the two countries. In December 2016, an open skies arrangement was concluded between the two nations, which removed all capacity restrictions between Australia and China for each country's airlines. Zhu et al. (2019a) report that in 2005, only Beijing, Shanghai, and Guangzhou had direct flights to Sydney and Melbourne and most Chinese travellers used Hong Kong, Singapore, and even Seoul as a transit point to Australia. However, in 2016, seven Chinese airlines served the China–Australia market, from China's ten cosmopolitan cities to Australia's major capital cities.

In 2016 China Southern was the largest contributor (38 percent) to the direct connectivity between China and Australia, followed by China Eastern's 21.8 percent and Air China's 18.6 percent. Qantas only made a contribution of 6.2 percent (Zhu et al., 2019a). Guangzhou has forged its status as a significant transfer hub between Australia and China, thanks to China Southern's contribution. China Southern has increased its flight routes to Australia from Guangzhou since 2009. In 2012, China Southern signed a strategic cooperation agreement with Tourism Australia to build the "Canton Route"—the route linking Europe, and Australia via Guangzhou. It has since then launched non-stop services to all major Australian capital cities from China, including Adelaide, Brisbane, and Perth. Apart from increasing frequency and destination to the Australia market, China Southern has worked with Guangzhou Immigration and Customs to simplify the transfer procedure and launched 'through check-in' service in 2012, which means that the transit passengers do not need to reclaim their baggage at Guangzhou airport for customs clearance, regardless of whether they are travelling out of or into China. In addition, China Southern provides transit passengers with free transit lounge services if the transit time exceeds four hours. In 2012, the number of transit passengers using the "through check-in" service increased by 458,000, while in 2016, this number increased to 1.74 million. Table 2.8 lists the annual flights between China and Australia by China's big three.

China's liberalisation efforts were recognised by IATA, which appreciated China's commitment to "progressively liberalising the aviation market and the

TABLE 2.8 Annual flights between China and Australia in selected years, 2007–2017

Airline	2007	2009	2011	2013	2015	2017
China Southern	620	652	2,476	3,236	3,428	4,800
Air China	696	1,048	1,400	1,412	1,576	2,000
China Eastern	704	720	1,192	1,948	2,084	3,528

Source: IATA AirportIS.

work to harmonise the air traffic control with global standards" (IATA, 2005). The approach was also welcomed by the USA, with the Secretary of the Treasury Henry M. Paulson Jr. (2008) commenting that China's growth and stability depended on moving forward with liberalisation, despite resistance from its domestic industry, rather than economic nationalism which would have an adverse impact on the economy of both the USA and China.

China's three strategic cities and the airport sector

Beijing, Guangzhou, and Shanghai, strategically located at the North, South, and East of the territory, are traditionally considered the most important cosmopolitan cities both politically and economically, which still holds true today. Beijing, the country's capital, has been an international gateway ever since China has had its international operations. Its close proximity to the Bohai Sea via Tianjin, a historical and modern industrial city focusing on aviation and aerospace, has created a Beijing–Tianjin Economic Zone and is one of the key economic drivers for Northern China. Guangzhou is at the far end of China's southeast coastline at the mouth of the Pearl River, with proximity to Hong Kong and Macau. Sharing a lot of cultural features, Guangzhou, together with Shenzhen and Zhuhai in the Pearl River Delta (PRD) region, evolved into a metropolitan interlocking region linking a network of cities (Williams, 2009), thus creating an unrivalled competitive edge for its economic growth. Being the first in establishing special economic zones and maintaining its robust economic growth over the past 30 years, Guangzhou and the PRD region boast to be the global workshop and centre of electronics and pharmaceutical products. Shanghai is the most commercialised city in China and has always been enjoying its prosperity as the country's business centre. Located at the mouth of the Yangtze River, the city, together with other major industrial cities in its catchment area, including Hangzhou, Nanjing, and Suzhou, has also formed a metropolitan region and has acted as one of China's economic powerhouses since 1978. In addition, as one of the biggest sea ports of the world, Shanghai is advantageous in being able to integrate all transport modes with the potential to be a real intermodal hub in the Asia-Pacific region.

Infrastructure including airports, highways, and railways has received magnificent amount of investment in China since 1978. According to CNSB, around

6.4 billion yuan (US$94 million) was invested in all transport modes in 1978 (US $181 Billion). In terms of air transport, about 131.1 million yuan (US$14.7 million) was invested in 1980 in airport construction. The figure went up to 10.6 billion yuan (US$1.2 billion) in 2003 and then 30 billion yuan (US$4.3 billion) in 2008. The investment brought more civil airports into operation. In 1980, there were 80 airports for civilian use with one airline, which was CAAC,[5] operating on 159 domestic routes. In 2007, there were 148 licensed civilian airports connecting 146 domestic cities with total passenger throughput of 387.6 million, representing 16.8 percent increase year on year, and a total freight throughput of 8.6 million tonnes, a 14.3 percent increase (Li & Fung Research Centre, 2009). The increased number of airports has helped to improve the density of airports to 1.6 per 100,000 square metres across China, though still quite low compared with 23.2 airports per 100,000 square metres in Japan. When considering the catchment area of an airport, being measured as 1.5 hour's drive time, China's air transport service only covers 62 percent of the population and 82 percent of its national GDP (Diao, 2010). The imbalance of airport distribution in China is also structured with most airports being in Eastern China while a few are in the western areas, with the density in Xinjiang and northwest China being only 0.6 and 0.8 airports per 100,000 square metres, respectively (Figure 2.5) (Diao, 2010). When comparing China with other countries such as the USA, Japan, France, India, and Australia, its average airport density is much lower (Figure 2.6) (Diao, 2010).

To improve the situation, in 2010, CAAC announced an investment of 250 billion yuan (US$35.7 billion) in airport constructions with an aim of bringing the total number of civil airports to 244 by 2020 (excluding those in Hong Kong, Macau and Taiwan), with the majority of the new airports being built in Western China. Well on its way to achieving this target, in 2017, there were 229

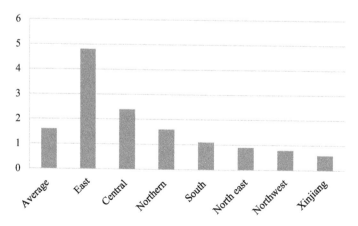

FIGURE 2.5 China's airport density per 100,000 square metres in 2009.
Source: Diao, (2010).

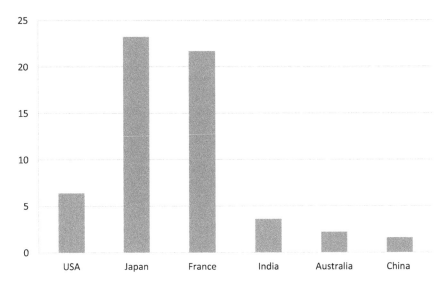

FIGURE 2.6 Comparison of airport density per 100,000 square metres between China and a few selected countries in 2009.

Source: Diao, 2010.

licensed civil aviation airports across the mainland, connecting 224 cities and 228 airports with scheduled services according to the CAAC. Passenger traffic in 2017 exceeded 2016 by 12.9 percent with 1.1 billion passengers carried in total, of which approximately one billion were transported on domestic routes. Cargo and mail totalled 16.17 million tonnes (7.1 percent higher than the previous year) of which ten million tonnes were on domestic routes. Considering that only 47 airports handled more than one million passengers in 2007, vis-à-vis the 32 airports with annual passenger traffic exceeding ten million in 2017 (representing 81 percent of total traffic), the rapid expansion in aviation across mainland China in the past decade comes into sharp focus. Nevertheless, comparatively, smaller airports in the market have maintained market share overtime relative to the overall growth of passenger demand in China. In this respect, it was revealed that the 26 airports with passenger share of 2–10 million accounted for 11.8 percent of passenger traffic. In total airports with passenger traffic over two million therefore accounted for 92.8 percent, relatively equitable to distribution in 2007 when larger airports were compared to the smaller facilities as noted prior.

Beijing Capital Airport remained champion of the league reporting 95.7 million passengers, followed by Shanghai Pudong with 70 million, and Guangzhou Baiyun with 65 million (Table 2.9). In terms of cargo, Shanghai Pudong ranked first, followed by Beijing Capital and Guangzhou Baiyun, which collectively handled 47 percent of the national total (Table 2.10). The most phenomenal growth comes

TABLE 2.9 Top 14 airports measured by passenger throughput in 2016 and 2017

Airport	Passenger throughput (million, per annum)			
	Rank	2017	2016	Change (%)
Beijing/Capital	1	95,786,296	94,393,454	1.5
Shanghai/Pudong	2	70,001,237	66,002,414	6.1
Guangzhou/Baiyun	3	65,806,977	59,732,147	10.2
Chengdu/Shuangliu	4	49,801,693	46,039,037	8.2
Shenzhen/Bao'an	5	45,610,651	41,975,090	8.7
Kunming/Changshui	6	44,727,691	41,980,339	6.5
Shanghai/Hongqiao	7	41,884,059	40,460,135	3.5
Xi'an/Xianyang	8	41,857,229	36,994,506	13.1
Chongqing/Jiangbei	9	38,715,210	35,888,819	7.9
Hangzhou/Xiaoshan	10	35,570,411	31,594,959	12.6
Nanjing/Lukou	11	25,822,936	22,357,998	15.5
Xiamen/Gaoqi	12	24,485,239	22,737,610	7.7
Zhengzhou/Xinzheng	13	24,299,073	20,763,217	17.0
Changsha/Huanghua	14	23,764,820	21,296,675	11.6

Source: www.caac.gov.cn.

TABLE 2.10 Top 14 airports measured by Cargo volume in 2016 and 2017

Airport	Freight (ton)			
	Rank	2017	2016	Change (%)
Shanghai/Pudong	1	3,824,279.9	3,440,279.7	11.2
Beijing/Capital	2	2,029,583.6	1,943,159.7	4.4
Guangzhou/Baiyun	3	1,780,423.1	1,652,214.9	7.8
Shenzhen/Bao'an	4	1,159,018.6	1,125,984.6	2.9
Chengdu/Shuangliu	5	642,872.0	611,590.7	5.1
Hangzhou/Xiaoshan	6	589,461.6	487,984.2	20.8
Zhengzhou/Xinzheng	7	502,714.8	456,708.8	10.1
Kunming/Changshui	8	418,033.6	382,854.3	9.2
Shanghai/Hongqiao	9	407,461.1	428,907.5	-5.0
Nanjing/Lukou	10	374,214.9	341,267.1	9.7
Chongqing/Jiangbei	11	366,278.3	361,091.0	1.4
Xiamen/Gaoqi	12	338,655.7	328,419.5	3.1
Tianjin/Binhai	13	268,283.5	237,085.2	13.2
Xi'an/Xianyang	14	259,872.5	233,779.0	11.2

Source: www.caac.gov.cn.

from those medium-sized airports such as Guoluo, Shihezi/Huayuan, Qionghai, Ninglang, and Liping which registered year-on-year growth exceeding 1,000 percent and even 20,000 percent in the case of Guoluo. Freight at these facilities increased ranging from 0.04–3.30 tonnes in 2016 to 11–110 tonnes in 2017.

FIGURE 2.7   Top 20 airports in 2017 measured by passenger throughput.
Source: www.aci.org.

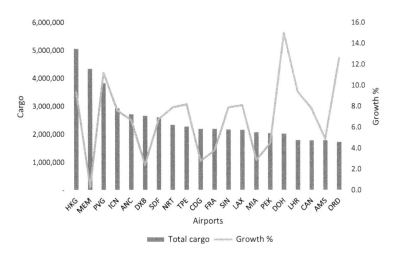

FIGURE 2.8   Top 20 airports in 2017 measured by cargo volume.
Source: www.aci.org.

The achievement was reflected in ACI's world's busiest airports league. In the 1980s and 1990s, none of the Chinese airports were listed in the world's top 30 airports in terms of passenger and cargo traffic. The first time that a Chinese airport appeared in the league was in 2002 when BCIA was ranked 26th, with a total passenger throughput of 27.2 million. In 2005, BCIA jumped to the 15th place with a record number of 41 million passengers. In 2017, mainland Chinese airports in the top 20 of league table included Beijing (2nd), Shanghai (9th), and Guangzhou (13th). Figures 2.7 and 2.8 show the top 20 airports in terms of passenger and cargo traffic, respectively, for 2017.

Improvement of China's airport connectivity

The concept of connectivity in air transport was first introduced to evaluate the importance of an airport in terms of its connection to other airports

(Zhu et al., 2018). Zhang et al. (2017), Zhu et al. (2018), and Zhu et al. (2019b) have developed connectivity measure to quantify an airport or a city's connections with other cities or countries. This measure not only considers quantity indicators, such as the number of seats, but also travel quality indicators, such as travel time, aircraft type, etc. This kind of measure can be used to evaluate the role of an airport in the existing air transport network, helping detect weak points and seek ways to improve the reliability and accessibility of the network to reduce travel time and costs (Hadas et al., 2017). Table 2.11 lists the air connectivity scores for the top 20 airports in China in selected years from 2006 to 2016. Note that like consumer price index, these connectivity scores are only meaningful when they are used to compare the level of connectivity across airports or over time. It can be seen that Chinese airports were tremendously successful in increasing their connectivity from 2006 to 2016. The increase is particularly impressive in secondary-tier cities (Beijing, Shanghai, and Guangzhou are normally regarded as the first-tier cities and most provincial capital cities are the second-tier ones).

Table 2.12 reports the major Chinese carriers' link connectivity at the domestic market level, which is the aggregation of their route-level connectivity. From 2007 to 2017, the connectivity of the big four carriers more than doubled in the domestic market, suggesting that China is a growth market. The performance of

TABLE 2.11 Airport connectivity, 2006–2016

Airports	Y2006	Y2008	Y2010	Y2012	Y2014	Y2016
Beijing Capital	5,943	6,608	7,243	7,909	8,489	8,762
Guangzhou	3,395	4,032	4,607	5,219	5,746	6,095
Shanghai Pudong	3,236	3,529	3,849	4,292	4,721	5,693
Kunming	1,845	2,194	2,730	2,957	4,001	4,805
Shenzhen	2,335	2,669	3,192	3,497	4,296	4,729
Chengdu	2,247	2,333	2,945	3,406	3,868	4,585
Xi'an	1,257	1,392	2,240	2,977	3,609	4,116
Shanghai Hongqiao	2,281	2,670	3,232	3,619	3,907	3,980
Chongqing	1,119	1,522	2,034	2,746	3,339	3,800
Hangzhou	1,302	1,683	2,045	2,362	2,960	3,565
Urumqi	769	845	1,499	2,079	2,508	3,071
Xiamen	997	1,279	1,734	2,260	2,789	2,965
Harbin	583	890	1,130	1,380	1,909	2,508
Nanjing	742	1,185	1,554	1,769	1,900	2,484
Zhengzhou	576	797	1,170	1,392	2,043	2,392
Qingdao	918	1,084	1,440	1,685	2,032	2,370
Changsha	851	1,150	1,583	1,818	2,063	2,320
Shenyang	789	1,030	1,226	1,436	1,822	2,225
Wuhan	724	1,081	1,368	1,578	1,930	2,200
Dalian	767	980	1,383	1,508	1,847	2,157

Source: Zhang et al. (2017).

TABLE 2.12 Airline connectivity in China's domestic market, 2007–2017

	China Southern	China Eastern	Air China	Hainan Airlines	Spring Airlines
2007	121,576	83,425	71,628	39,800	1,113
2009	140,861	105,151	86,904	45,442	7,296
2011	152,786	118,108	96,091	46,769	13,349
2013	174,343	140,976	106,542	60,453	18,460
2015	193,412	160,837	115,796	64,430	20,276
2017	200,750	179,384	126,263	80,379	25,963

Source: Calculated by the authors based on IATA AirportIS' historical schedule data.

Spring Airlines, an LCC, experienced a 23-fold increase in the domestic market. However, compared with the big three, Spring is still small in scale given that its connectivity was only 13 percent of China Southern's in 2017.

There is no doubt that China will overtake the USA to be the largest aviation market in the near future, thanks to the continuous demand for air travel. Deregulation or liberalisation measures after 2003 have contributed to this likely outcome. The liberalisation moves include the introduction of private capital into China's air transport, leading to the two waves of the establishment of private carriers and LCCs in 2005 and 2013, respectively. These new entrants, together with the emergence of HSR, put much competitive pressure on the air transport industry and create momentum for Chinese carriers to seek improvement in efficiency and competitiveness as well as new markets. In the meantime, China has taken a different attitude towards the liberalisation in its international air transport. Liberal arrangements have been made with some major markets including the USA, ASEAN, Korea, Japan, etc. These open and liberal arraignments have given the Chinese carriers the room to grow and the chance to become stronger. They in turn call for more liberalisation moves to allow them to participate in the provision of more international air transport services.

Latest development

However, the recent trade disputes between China and the USA, and Brexit have cast uncertainty over the air transport sector. Worldwide demand for air freight fell by 4.7 in April 2019 compared with the same period in the previous year, and a larger fall was recorded among the Asia Pacific region, according to the IATA data. Law et al. (2018) also notice that for decades the US carriers were firm supporters for air transport liberalisation, but recently they asked the government to end open skies agreements with Qatar and United Arab Emirates to stop allowing the Gulf carriers to expand in the US market arguing that the gulf carriers received subsidies from their government and competed unfairly on the transatlantic route. Similar voice was also expressed in Europe. It would be interesting to keep a close eye on how the rise of global protectionism impacts China's air transport liberalisation process, which is possibly a new research topic worth examining. Just as these varied shifts in the geopolitical and more pointedly, the aero-political landscape are not without consequence for air transport in China, associations between China's policies to strengthen its own geopolitical stance and its air transport industry are also noteworthy. From a geopolitical strategy perspective, this analysis would therefore not be complete without musings on its signature contemporary geopolitical strategy.

Notes

1 Dr Yahua Zhang is an Associate Professor of Economics at University of Southern Queensland. His research interests include aviation economics, industrial organisation and transport logistics. Dr Zhang has published extensively in leading transport and economics journals, including Transportation Research Part A: Policy and Practice, Journal of Transport Geography, Journal of Transport Economics and Policy, Transport policy, Journal of Air Transport Management, International Journal of Industrial Organisation, Review of Industrial Organisation, etc.
2 In 2009, China overtook Japan to become the world's second largest economy.
3 Great Wall International Airlines is a joint venture set up by China's Great Wall Industrial Co Ltd and Singapore Airlines.
4 Jade Cargo International is a joint venture set up by Shenzhen Airlines and Lufthansa Cargo.
5 Until early 1990s, CAAC was both a regulator and an air carrier.

References

CAAC, (2008), Statistical data on Civil Aviation of China, 2008, Civil Aviation Press of China, Beijing.

CAAC, (2007), CAAC's Policy Proposal with respect to the Party's Principles of the Sixth Plenary Session of Sixteenth Congress, July 2007; available at http://www.caac.gov.cn/B7/200707/t20070719_6569.html

Chen Z., (2017), Impacts of high-speed rail on domestic air transportation in China. *Journal of Transport Geography*, 62, 184–196.

Diao Y., (2010), China airport construction and new development, Trends in the next decade, in: China Civil Aviation Development Forum 2010, 12–13 May 2010, Beijing, China

Doganis R., (2001), *The airline industry in the 21st century*, (1st edition). Routledge, London.

Forsyth P., King J., and Rodolfo L., (2006), Open skies in ASEAN, Journal of Air Transport Management, Vol. 12, 2006, pp 143–152

Freeman P. and Lam R., (2002), Current Chinese Aviation Policy in the Light of Economic Globalisation, 2002, Air and Space Law, Feb. 2002

Fu X., Lei Z., Wang K., and Yan J., (2015), Low cost carrier competition and route entry in an emerging but regulated aviation market–the case of China. *Transportation Research Part A: Policy and Practice*, 79, 3–16.

Hadas Y., Gnecco G., and Sanguineti M., (2017), An approach to transportation network analysis via transferable utility games. *Transportation Research Part B: Methodological*, 105, 120–143.

IATA, (2005), IATA honours Chinese Minister with gala award, IATA press releases, on 30th May 2005, available at http://www.iata.org/pressroom/pr/Pages/2005-05-30-03.aspx

ICAO, (2013), ICAO working paper: expansion of market access for international air transport in a proactive, progressive, orderly and safeguarded manner, available at www.icao.int/Meetings/atconf6/Documents/WorkingPapers/ATConf.6.WP.97.2.1.en.pdf

Li & Fung Research Centre, (2009), China Distribution & Trading, January 2009, Issue 57,

Law C. C., Zhang Y., and Zhang A., (2018), Regulatory changes in international air transport and their impact on tourism development in Asia Pacific, in: Fu, X. and Peoples, J., (editors), *Airline Economics in Asia* (pp. 123–144), Emerald Publishing Limited, Bingley.

Li H., Strauss J., and Lu L., (2019), The impact of high-speed rail on civil aviation in China. *Transport Policy*, 74, 187–200.

Liu, S., & Oum, T.H, 2018, Prospects for air policy liberalization in China as a Result of China-ASEAN Open skies: Changing role of Chinese mega carriers in global scene and anticipated Low Cost Carrier competition, *Transport Policy*, 72(C), 1–9.

Ma W., Wang Q., Yang H., and Zhang, Y., (2019), Is multimarket contact an antitrust concern: a case of China's airline market. *Transportation Research Part A: Policy and Practice*, 132 (C), 515–526.

Ma, W., Wang, Q., Yang, H., Zhang, A., & Zhang, Y. (2019). Effects of Beijing-Shanghai high-speed rail on air travel: Passenger types, airline groups and tacit collusion. *Research in Transportation Economics,* 74, 64–76.

National Bureau of Statistics, (2008), *China statistical yearbook 2008*, China Statistical Press, Beijing.

National Bureau of Statistics, (2018), *China statistical yearbook 2018*, China Statistical Press, Beijing.

Oum T. H., and Lee, Y. H., (2002), The Northeast Asian air transport network: is there a possibility of creating Open skies in the region? *Journal of Air Transport Management*, 8(5), 325–337.

Pagliari R., (2010), Trends in air service development within the highlands and Islands of Scotland 1983–2006, in Williams, G. and Brathen, S., (editors), *Air transport provision in remote regions*, (pp. 21–46) Ashgate Publishing, Cornwall, UK.

Paulson H.M. Jr., (2008), The Right Way to Engage China, Strengthening US-China Ties, Foreign Affairs, Sept/Oct Issue

Xia C., (2014), Understanding the development of China's private airlines. 30–2014–0482, available at http://news.carnoc.com/list/296/296880.html

Wang J., and Jin F., (2007), China's air passenger transport: An analysis of recent trends, *Euroasian Geography and Economics*, 48 (4): 469–480, Bellwether Publishing, Ltd.

Williams A., (2009), Contemporary Issues Shaping China's Civil Aviation Policy, Balancing International with Domestic Priorities, Ashgate Publishing Ltd, Farnham

Yu H., Zhang Y., Zhang A., and Wang K., (2019), A comparative study of airline efficiency in China and India: A dynamic network DEA approach. Unpublished manuscript, University of Southern Queensland.

Zhang A., Wan Y., and Yang H., (2019), Impacts of high-speed rail on airlines, airports and regional economies: A survey of recent research. *Transport Policy*, 81, A1–A19..

Zhang Y., and Round D. K., (2008), China's airline deregulation since 1997 and the driving forces behind the 2002 airline consolidations. *Journal of Air Transport Management*, 14(3), 130–142.

Zhang Y., and Zhang A., (2016), Determinants of air passenger flows in China and gravity model: deregulation, LCCs, and high-speed rail. *Journal of Transport Economics and Policy (JTEP)*, 50(3), 287–303.

Zhang Y., and Zhang A., (2017), Air transport development: a comparative analysis of China and India, in Finger, M., and Button, K., (editors), *Air transport liberalization* (pp. 112–137) Edward Elgar Publishing, Glos.

Zhang Y., Zhang A., Zhu Z., and Wang K., (2017), Connectivity at Chinese airports: the evolution and drivers. *Transportation Research Part A: Policy and Practice*, 103, 490–508.

Zhang, Y. and Lu Z., (2013), Low cost carriers in China and its contribution to passenger traffic flow, *Journal of China Tourism Research*, 9, 207–217.

Zhou W. and Szyliowicz J., (2006), *World Transport Policy & Practice*, Volume 12, Number 4. Eco-Logical Ltd., Lancaster.

Zhu Z., Zhang A., and Zhang Y., (2018), Connectivity of intercity passenger transportation in China: A multi-modal and network approach. *Journal of Transport Geography*, 71, 263–276.

Zhu Z., Zhang A., Zhang Y., Huang Z., and Xu S., (2019a), Measuring air connectivity between China and Australia. *Journal of Transport Geography*, 74, 359–370.

Zhu Z., Zhang A., and Zhang Y., (2019b), Measuring multi-modal connections and connectivity radiations of transport infrastructure in China. *Transportmetrica A*, 15(2), 1762–1790.

3

ANALYTICAL FRAMEWORK

Applying the Micro-Macro Linkage Approach

International trade policymaking

International air transport exemplifies the features of international trade but with a higher degree of political involvement than other trade sectors. International air transport policy is integrally related to a nation's overall trade policy. This is evident considering the extent to which aviation plays an important role in the global economy, supporting and facilitating trade, investment, and the movement of goods and people. Trade policy is defined by Cohen (1968) as the sum total of actions by the state intended to affect the extent, composition, and direction of its imports and exports of goods and services. It deals with the economic effects of direct or indirect government intervention that alters the environment under which international transactions take place (Kerr, 2007).

Since the late 1970s, when policies started to shift towards neoliberalism, economics has become as important as security in international relations resulting in international trade policy becoming increasingly interlinked with a nation's foreign policy. To this end, the consequential impact on other issues has also increased and China's aviation industry provides a testimony and basis for analysis. Economic growth via international trade has become a priority on the agenda of a government's national strategy, with economic objectives always being reflected in a nation state's foreign policy. As Alexis Johnson, Undersecretary of State of the USA argued, economic considerations would dominate foreign policy over the following decade as security concerns had dominated the previous two decades (Johnson, 1972). The prioritised significance of economics in foreign policy matters being perpetuated across the global arena added complexity to managing existing delicately yoked national objectives. In other words, governments increasingly had to ensure that political pursuits and objectives and economic strategies were not conflictual.

The increased exchange of economic activities and rapid growth of international trade across borders brings about an interdependent relationship between nations. Governments are seen more and more often sitting at tables negotiating for trade agreements, trying to secure a deal that would best satisfy their respective needs to protect while at the same time optimise their national interests. However, because of the complicated and multi-faceted nature of international trade issues and the sophisticated negotiations involving another country which may or may not share any identical political, social, economic, and cultural features, governments and their institutions tend to find themselves struggling between powers, benefits, interests, and other considerations. Although increased economic interdependence helps the advance of globalisation, it adversely makes the international trade policymaking much more complicated, bringing in more issues and actors for consideration in the process (Bayne and Woolcock, 2007).

These phenomena have attracted unprecedented attention from both academia and industry practitioners who have attempted to address the subject matter from both an economic and political perspective with a wealth of literature being produced. Economists take it as an extension of domestic economic activity, trying to understand how international economic activities could help to maximise domestic economic objectives, while political scientists treat it as a political game played by various stakeholders, thus focusing the analysis on the politics involved in the policymaking process.

International relation theories are borrowed to examine the bilateral diplomatic and military relations which are considered fundamental in formulating foreign trade policy. Different theories have been developed with the aim of analysing the different roles played by different factors involved in the process of policymaking, with a view to attempting to deepen the understanding as to how and why governments and other actors in the process of trade policy making and negotiation behave in the way they do. As the subsequent review reveals, existing approaches have largely been predicated on two analytical predilections rooted in economic and political underpinnings. However, critical reviews have proven to be too parochial in this context. This has fuelled the need to use a more comprehensive framework given the ubiquitous influences exerted on the aviation industry as expressed prior, coupled with China's uniqueness. The remainder of the chapter presents a roadmap towards the validity of a more comprehensive framework for this analysis and then applies that to China's aviation environment.

A critical review of the analytical approaches to international trade policy

There have been two primary schools of thought as influenced by both economists and political scientists to assess the various interactions and the emergence of consensus in the formulation of trade policy. The various tools, models, and divergent theories—each of which can be legitimated proved and supported—assess interactions among various factors forming the policy. However, policymaking by nature is an isolated innate yet multifaceted process. Trade policy interfaces increasingly with science, such as food safety and environment, and the analysis of consumer preferences, such as animal welfare and child labour, as well as social policy, such as sustainable development and labour standards. In many cases, changes in trade policy have been found to implicate domestic vested interests to the benefit of foreign competitors.

Economic perspective

Economists tend to analyse the making of economic policy through a normative approach, modelling policy choices as the equilibrium outcome of a well-specified strategic interaction among rational individuals (Persson and Tabelline, 2000), who act and react within the institutional context which shapes their behaviours (Muscatelli, 1996). Rational policymakers can define clearly the problems, identifying the aims, considering the alternatives, developing clear criteria for choice, and monitoring the decisions (Dumbrell, 1997). Within the ambit of perceived rationality, it often reasonably ensues that these policymakers seek optimisation in achieving their personal objectives for the benefit of their own welfare and self-interest. Consequently, the extent to which these interests will coincide with that of the wider community is often open to debate. In this pursuit, they will follow their personal preferences for the goods and services available to seek to be elected or re-elected. However, when operating in a world which is not ideal but full of acute difficulties, ranging from unreliable information to time and electoral pressures, policymakers have to make choices on the basis of picking the least unsatisfactory option as a result of various constraints.

The potential conflict between individual and social objectives leads to several important and related questions which economists struggle to address. For example, Muscatelli (1996) challenged the rational model with the following questions: "If there is a conflict which emerges between the pursuit of self-interest on the part of political and economic institutions and the wider community, what mechanisms can be devised to ensure the society's welfare is maximised?" "What are the potential conflicts between the pursuit of economic welfare and other social and political objectives, such as democracy and social justice?" A variety of models have been developed to explain the determination of domestic and international politics within this economic self-interest framework (Baldwin, 1996). The analytical tools used by economists working to address this type of question are by no means novel, with Muscatelli (1996) observing that the use of game-theoretical analysis was reminiscent of similar applications in the areas of welfare economics and industrial economics.

Although the rational choice model is useful in giving explanations and insights into making choices within the bounded alternatives available, it is not short of criticism. There are certain other factors which are involved in international economic policymaking for which economics fails to offer any answer, for example, national security concerns and environmental considerations (Dam, 2001). Dam further pointed out that most of the public disputes over an economic approach to international economic issues did not revolve around non-economic values, but rather were a by-product of the political system through which international economic policy decisions were actually made. For example, the metrics that economics used to calculate international economic issues leading to a policy result were regarded as controversial.

Another drawback is the neglect of the role played by institutions and ideology in shaping trade policy. Institutions, as Muscatelli (1996) claims, are not only exogenous constraints restricting the economic behaviour of policymakers, but able to respond to economic phenomena to reflect the changing context as a two-way interaction. Goldstein (1988) sees institutions as the embodiment of prevailing policy ideas which, once created, ossify, though sometimes enduring long after the ideas which give rise to them have lost favour. She describes the institutions as the annual sediment on a flood plain, slowly building up but powerfully influencing the river flow. Levine (1994) claims that it is the ideas and attitudes of economic decision makers, whether simple and intuitive or highly complex and formally articulated, that structures their environment for choice, informs their consideration of various courses of action, and provides rationalisations for the choices that are made. Baldwin (1996) contends the ways that economists apply the economic approach universally to their analysis of international economic policymaking by presuming that every country operates within identical political and social circumstances, which, in reality, are dynamic and divergent. Certainly, in the context of China, this postulate would be an unstable foundation from which to launch an analysis of China's aviation environment given its unquestionable "uniqueness." Another weakness of the economic perspective is its ignorance of the changing nature of society and environment over time. Thus, the lack of appreciation of such changes causes its failure to explain the evolving factors that shape the economic policy. Once again, the reality is that China as a society has undergone various changes over time and the implications of actions resulting in gradual liberalisations across the society cannot be underestimated nor conflated with implications which may have emerged in other societies globally.

Suettinger (2003), with his personal experience in serving various senior positions in the US government, casts significant doubts on the rational model of foreign policymaking. He believes that the process of foreign policymaking is a "strategic analysis on both sides of the goals and intentions of the other." He asserts:

> first, foreign policies are not the product of pristine calculations of national interests by trained experts with all the facts at their disposal. Rather, politics are the result of a profoundly political process in which differing, sometimes competing, domestic interests, bureaucracies and individuals affect the outcome. Although some of the key players are well-informed

experts, they are often working with incorrect or incomplete information, as well as inaccurate assumptions and cultural prejudices. Second, strategic assessments that extrapolate historical or ideological trends and project future policies and behaviours are likely to be wrong, as they seldom take account of the domestic politics of decision-making or the effect of unpredictable events that often drive the process.

(Suettinger, 2003)

In an attempt to understand the mindset of economists, who acknowledge that politics is important and that foreign trade policy represents the interaction of politics and economics but still leaves political analysis aside, Pastor (1980) argues that economists tend to stress the distinctiveness and incompatibility of political and economic analysis which is not helpful in explaining the political process or describing the foreign economic policy in the comprehensive and systematic way that would permit one to draw conclusions about the political causes or consequences. Consequently, no guide or conceptual framework is available to help understand the politics of policies as diverse as foreign trade policies and to state a proposition about how it is made.

The political perspective

Likewise, political scientists attempt to address policymaking from a political perspective, believing that policymaking is an inherently political process that centres on perceptions, value judgements, the setting of priorities, the making of choices, and the distribution of choices (Cohen, 2000). Cohen argues that no economic policies are made that are 100 percent free of political overtones which try to strike a balance of equity versus efficiency. Whether to emphasise social fairness or economic efficiency raises the very basic question regarding to what extent government intervention in international economic relations is desirable. When both national security and domestic economic well-being, which are two separate realms, are convened to be the goals that the national government wishes to achieve through international economic activities, considerations of optimising the national economic strength will take priority to become the central component in world politics, since economic strength is widely recognised as having become part of a broader definition of national security (Cohen, 2000). The pursuit of such are not necessarily complementary internal and external goals creating a situation whereby the international trade policymaking process is characterised with a juggling act that reconciles conflicting domestic political pressures, domestic economic policy objectives, and foreign policy priorities (Cohen, 2000). Politics, therefore, is played in the whole process where influences can be exerted, thus generating a heated discussion as to what factors would play a bigger role. Ikenberry et al. (1988) have summarised the various approaches into three levels that are of more significance: system-centred, societal-centred, and state-centred approach.

System-centred approach

The system-centred approach, or international approach, is supported by quite a few influential theories which concentrate on the function of attributes or capabilities of one country relative to other nation states (Ikenberry et al., 1988), with government officials perceived as responding to particular sets of opportunities and constraints that a nation's position in the international system creates at any moment in time. The international system is a necessary "first cut" in any analysis of international or comparative politics, as well as being important in the study of foreign trade policy in a single nation state, as it is able to explain recurring international events and the commonalities in national foreign policies.

One of the most famous theories is that of hegemonic stability which holds that the existence of hegemonic power is a necessary condition for the existence of international transactions. Gilpin (2001) states that a hegemon possesses two dimensions of power: political and military strength as well as economic efficiency. With this power, the hegemon has the resources to force or induce others to adopt liberal practices in their foreign trade, thus being able to create and maintain a stable international economic order which decisively influences the other country to participate. Being considered to be subordinate to politics, economics is used to optimise the national interest that enables the state to take actions to achieve its goals of security, welfare, and other societal values. Although economic welfare is pursued for its own sake, the ultimate objective is to be instrumental for political power. The analysis is thus focused on the power distribution among states within the international system, with international actions being regarded as a product of global structure, defined by the distribution of power and resources (Lipson and Cohen, 2002). By understanding a state's sources of strength and areas of vulnerability in relation to other states, a better understanding of the creation of foreign trade policy is consequently developed.

The hegemon theory has been primarily used to account for the role of the USA in its creation and maintenance of an international trade regime, and in particular, to explain how the international economic order was established after World War II under US hegemony. In the case of international trade, it is used to explain why the USA prefers to pursue bilateral free trade agreements in which it is able to benefit from an asymmetric power relationship in its favour, rather than multilateral arrangements (Bayne and Woolcock, 2007).

However, as the theory only identifies the international constraints placed on nation states without considering the domestic political process, it is limited to explaining recurrent patterns of behaviour within the international arena and is inadequate in explaining the foreign trade policy in a single country (Ikenberry et al., 1988). It also fails to describe how a free trade regime is established, maintained, and abandoned, with Stein (1990) arguing that a hegemon cannot bring about an open trading order alone without getting others' agreement. Trade liberalisation, therefore, among major trading states is rather the product of tariff bargaining, where the hegemon is likely to be required to make important concessions to achieve the political objectives (Stein, 1990). Another weakness, as identified by Woolcock (2007), is that it is hard to measure power in the case of international trade. He argues that since the whole trade regime is built on the concept of reciprocity, countries have been reluctant to liberalise their markets unless they have been confident that other countries want to do so in an equal measure. When the relative market size provides a fairly good proxy for power, which lends more leverage to negotiations, the challenge lies when negotiations are shaped by other elements of power, such as national security considerations.

Regime theory argues that states would only cooperate when there are cross-border economic activities that require rules or norms of behaviour, such as transport and telecommunications (Woolcock, 2007). When applying the theory to analyse a nation's policy of international air transport, it can be used to explain why the two countries are willing to sit together with an attempt of striking an agreement with respect to international operations between the country-pair market, due to the fact that the parties concerned are signatories to International Civil Aviation Organization (ICAO) and are obliged to abide by the provisions of international conventions. However, it will not be appropriate to be used to explain why one party is willing to accept and agree what is requested and demanded by the other party, both of which are likely to have divergent agenda and objectives to achieve. One party is able to assert pressure for a more liberalised transport agreement but will find it hard to force the party at the other end of the country-pair market to accept what is proposed to its own benefit. When member states commit to binding obligations within ICAO's setting, they also limit the range of policy options open to them.

Society-centred approach

A second level of analysis is the society-centred approach. In contrast to the international level of analysis, emphasis is given to domestic politics either reflecting the preferences of the dominant group or class in society, or resulting from the struggle for influence that takes place among various interest groups or political parties. This approach views the policymaking process as a function of the interplay between organised societal interests and political institutions, thus resulting in government policy as being the outcome of a competitive struggle among affected groups for influence over particular policy decisions and the objectives being reflected in politicians' responses to different interest groups' demands (Bayne and Woolcock, 2007). Government institutions essentially provide a platform for group competition and do not exert a significant impact on the decisions that emerge (Ikenberry et al., 1988). This approach is widely applied to analyse the effects of interest groups on economic policymaking in the USA, which are primarily concerned with maximising the economic welfare of the individuals they represent. For example, Dam (2001) observes that the USA has constructed a political system with many points of access for interest groups and a number of elements to facilitate interest groups' influence. He further explains that one such element involves how the separation of power works to prevent a US president from taking action on an international economic policy without the support or the explicit approval of the Congress. Another point of access for interest groups is provided by the executive branch departmental structure, within which many departments and bureaus see their role in large measure as that of advocates for economic groups. With so many ways of influencing and places to access the policymakers, interest groups can frequently obtain favourable action, block unfavourable action, or redirect action by manoeuvres in the congressional committee system or in the labyrinth of executive branch departments and bureaus.

However, this approach, as Dam (2001) notes, is not necessarily applicable to other non-democratic systems where there is no platform for interest groups to get their voices heard. It is also criticised for a lack of theoretical rigour as it lacks a mechanism to measure independently the weight of group power, thus creating problems when identifying the dominant group or coalition at any time (Ikenberry et al., 1988). Both of these circumstances are particularly relevant given China's political system as explained. Its socialist political system though not necessarily oppressive does not embody the same structure as other political systems globally such as that of the USA.

In terms of international air transport, this approach is useful in analysing how different interest groups interact and how the outcome of the negotiation is influenced by those who either share common interests or hold competing views. Even within one industry, corporations of different sizes and business models tend to have divergent interests which all require satisfying by the policymakers, who are supposed to be open and unbiased in assessing the different views. This, in turn, has an impact on the negotiation outcome regarding to what extent a more liberalised air transport arrangement should be agreed upon, and whether protectionist measures should be in place to safeguard domestic businesses and markets. However, in applying this approach to international air transport, it should be noted that interest groups operate differently in different social systems. In countries like China with a non-democratic regime, representation of interest groups of the aviation industry is constrained by the system, whose approach of exerting impact on the policymaking process is, therefore, different from that in democratic countries such as the USA.

State-centred approach

A third level of analysis, state-centred explanation, stresses the significance of a state's institutional structure in shaping international trade policies (Baldwin, 1996). This approach has two focuses: one is to perceive the state primarily as an organisational structure, or a set of laws and institutional arrangements shaped by previous events, while the other is to assume the state to be an actor, concentrating on politicians and civil servants whose behaviour is to respond to internal and external constraints in an effort to manipulate policy outcomes in accordance with their personal preferences (Ikenberry et al., 1988). The institutional focus sees institutional change as nonlinear which occurs primarily at moments of significant crisis, while the individual focus argues that policymakers will intervene in the process in pursuit of objectives that are determined independently from domestic interest groups' narrow self-interested concerns. Although the state is the basic actor in international relations, state action can be analysed most effectively by concentrating on the behaviour of those individuals with a determining impact on international policies whose responsibility is to act for the state (Edwards and Wayne, 2009). Personal power and behaviour is critical in determining the direction of a nation's international trade policymaking.

Borrowed from other academic fields such as psychology and sociology (Abe, 1999) in analysing the behaviour of politicians and civil servants, this approach also focuses on cultural, behavioural, and psychological aspects of the individuals who are able to build new institutions to alter the distribution of power within government to achieve a specific goal, and are capable of mobilising inactive societal groups into the policy arena to offset their political adversaries while complementing their own interests (Ikenberry et al., 1988). Policymakers are shaped by ideas and beliefs and crafted by cultures where they are immersed. They have strong views on what constitutes a correct public policy and believe it is their duty to execute such policies, even in the face of contrary pressure from special interests (Zampetti, 2006). For example, the driving ideology of US foreign economic policy has been and remains the ideology of liberalism, advocating free trade, and optimisation of personal benefits (Dumbrell, 1997). The importance of ideas is highlighted by Zampetti (2006) who argues that ideas are able to fulfil the constitutive function for international society as well as shape the identity of states, which inspires policymakers to communicate and persuade their international counterparts to agree to coordination and cooperation in a dynamic and ever-changing international environment, where the identification of national interests is increasingly difficult due to globalisation and interdependence.

However, Woolcock (2007) argues that the role of individuals in the 21st century's international economic activities should not be overestimated to such an extent that an individual has mastery of the issues to sway over negotiations, due to the large number of actors being involved and the complexity of the issues. The individual-focused approach is also under scrutiny as it fails to address the international environment, or the policy effect of foreign governments, international organisations, and other foreign interest groups which have an impact on the considerations of individual policymakers. When applying this approach to international air transport policymaking of a particular country, it can be argued that its bureaucratic system carries specific features which are unique to reflect its social and economic system, hence shaping the corporate culture that in turn has an impact on its employees. It can also be argued that officials working for bureaucracies are subject to the influence of their personal background and experience. For example, the Civil Aviation Administration of China (CAAC) has developed out of a branch of the country's Air Force. Employees used to be quasi-military servants with military rankings who are characterised with obeying and following instructions. However, it is not convincing to assert that these officials working for the CAAC are empowered to dictate the policy objectives and negotiation outcome without taking into account their concerns of the overarching national political objectives.

Need for a comprehensive analytical framework

There is no doubt that the various approaches to examining trade policymaking have made a substantial contribution to an enhanced understanding of the field. The economic model is very helpful in understanding certain types of economic policymaking behaviour but is inadequate for analysing the full range of economic policies in a nation state and for undertaking comparative studies across countries. Political scientists have shed insight into some policy behaviour for which economic models failed to develop the subject from an all-embracing perspective allowing the examination of changes over time and across a country. All approaches, though, regardless of their focuses, international constraints, domestic determinants, or institutional and individual influences, can be more or less regarded as representing a single-level analysis (Zhao, 1996), all of which serve as the primary inputs leading to the formulation of the international economic policy as an output. By highlighting one attribute, Zhao (1996) argues that a single-level approach is valuable in explaining one aspect of the process of international policymaking, but fails to address other attributes which function simultaneously in this process. As a result of this sophistication, it is obvious that there is no single theory of any discipline that can provide definite answers on how states, under given circumstances, will conduct policy (Bayne and Woolcock, 2007).

Baldwin (1996), therefore, has called for a truly general framework to be developed which can be applied to the analysis so as to explain not only how the distribution of international economic and political power, domestic determinants, institutions, culture, values, and ideologies contribute to affect a country's international trade policies, but also how a country's trade policies have changed over time. Odell (1990) emphasises the need for synthesising the different approaches, noting that each of the models individually has proved inadequate as a single unifying vehicle. He, therefore, would welcome what he believes as an integrated theory of trade policy formulation emerging in a form that encompasses individuals, groups, and states as actors. Bayne and Woolcock (2007) have called for an analytical framework to help sort out the complex factors that influence the complicated and sophisticated decision-making process and to identify the main explanatory factors so as to enable some generalisation in the field of trade policymaking.

Great efforts have been made in attempts to synthesise the different models so that a linkage approach can be developed to allow all variables to be examined with the aim of gaining insight into the complicated nature of the subject matter for a comprehensive understanding and interpretation of the policymaking process. Baldwin (1996) has proposed a framework by bringing together the four major sets of actors he has identified whose interactions would determine a country's international trade policy, which include home government; individual citizens; common interest groups; and foreign governments, along with various other foreign groups such as international organisations, and other foreign-interest organisations. Among these actors, he further explains, the national government is the key participant in the process, since it makes the final policy decisions and implements the policies, while the other three actors, though they take action, only influence the nature of these decisions by exerting various forms of political pressure on the domestic government. He stresses that it is necessary to recognise the importance of these diverse factors as identified, each of which deserves full attention for a rigorous analysis and examination.

Baldwin (1996) further argues that an even broader framework is needed when international trade policies are being analysed in places and times with different political and economic institutions and ideologies from those that currently exist in the USA, because the different regimes have different systems which will only work in certain circumstances. For example, the factor of common interest groups will have little or no influence on the officials in a non-democratic government who are not seeking election or re-election, while bureaucratic institutions and domestic political regimes are likely to outweigh the rest of the factors in deciding what the government wants to achieve through international economic activities. Although Baldwin proposes an analytical framework, he himself does not apply it to any empirical research.

Approaches applied to analyse China's foreign policymaking and its international behaviour

According to Hao and Su (2005), early studies of Chinese foreign policy are based on the state-centric assumption of a traditional realist approach to the study of international politics, treating policymaking as the product of a rational and unitary state that pursues and maximises its national interests under the constraints imposed by the external environment. This is because China is vulnerable to superpower dominance and manipulation, so that foreign policymaking is a reaction to such constraints, when the country tries to secure economic development objectives as well as strategic interests (Tow, 1994). Historical development in China gave way to culture and ideology having significant influences as the guiding principles of Chinese policymakers.

Since the 1990s, scholars studying China's foreign policymaking have adopted a societal factor approach, exploring the roles played by various social forces in the process of policymaking. These factors include the domestic political system, bureaucratic politics, mass media, decentralised local governments, and non-governmental players, which are all identified as internal sources in affecting policy formulation and change. Recognising the importance of domestic forces, this approach indicates that China's foreign policy is developed with the aim of addressing domestic pressures to reflect the concerns of the public. Interest group politics exists and plays in a different manner despite the absence of a democratic system.

Researchers such as Zweig (2002) have examined the role of the international environment and have argued that China's foreign trade policy is a response to international pressures. For example, import quotas on China's textile exports under the Multi-Fibre Agreement and surplus capacity in the global shipbuilding industry forced China to reform both sectors, and in the case of textiles, to shift to higher value-added products (Zweig, 2002). However, Zweig (2002) has noted that an explanation solely relying on international structure and pressures is insufficient, instead, domestic demand should be addressed which would help allow regulatory actions to be taken into account that facilitate the change. The analysis should be able to reflect the role of external forces, bureaucratic agents, domestic structure and a feedback loop based on the distributional consequences of internationalisation (Zweig, 2002).

The international–domestic linkage approach is adopted by Frieman (1994) who endeavours to understand China's foreign behaviour. She observes that the effects of the international system on Chinese foreign policy, citing international science and technology as an example, are really derivatives of its effects on domestic policy, with domestic policies being the driving force for the foreign policy, rather than vice versa. However, she explains that one could not simply categorise the set of international factors as "inputs," which, after being processed in China's policymaking machinery, would automatically lead to "outputs." This straightforward analysis only works to a limited degree and at a most general level. When making this calculation, one needs to appreciate how the machinery works to produce the "outputs." Examination of the full spectrum of implications for Chinese domestic policies should be pursued rather than looking for cause and effect between the international arena and Chinese foreign policy (Frieman, 1994).

Analysis that integrates the factors at three levels has been conducted by scholars including Shambaugh (1994), Zhao (1996), Lampton (2001), and Ng-Quinn (2004) who examined China's foreign policymaking process. Shambaugh (1994) analyses the China–USA relationship since 1949, exploring the patterns of interactions of factors at global, societal, and governmental levels. He asserts that the three levels analysis is very useful for considering complex relationships and ordering of multiple factors, but not necessarily helpful in examining other issues on the bilateral agenda such as Taiwan. It is valuable to explore the broad dynamics and patterns of interaction between the two complex, ambivalent, and contentious relationships. Lampton (2001) applies the three-level analysis on Sino-US relations between 1989 and 2000 which allows him to examine the impact of global institutions, domestic politics, mass media, and individual elite leaders on China's policy towards the USA. The approach enables him to depict a comprehensive picture of the Sino-US relations and how the two countries manage their relationship during that historical period of time, namely, after the June Fourth event and before China's entry into the WTO.

Ng-Quinn (2004) analysed the role of Chinese culture, idiosyncrasy in determining the perception of Chinese leaders, domestic political conflicts, and the international system, which works collectively to shape the formulation of Chinese foreign policy. He argues that factors at all levels should not be treated as a laundry list without giving any item special or prioritised attention, as Chinese foreign policy behaviour is foremost influenced by the external constraints superimposed by the structure of the international system. All the rest—at domestic, societal, and individual levels, including decision makers, domestic politics, as well as accidents, unintended consequences, coincidences, confusions, stupidity, etc.—are only relevant to the extent that they cause changes in Chinese capabilities, leading to changes in the distribution of power, and thus, structural transformation.

Yet, attempts at theorising linkage politics seems to have been abandoned because of the excessive complexities of the research object, with Abe (1999) observing that even analysing a state's domestic politics is complicated enough. When attempts are made to integrate domestic politics with international politics, the task becomes hopelessly difficult (Abe, 1999). Despite the efforts of trying to understand China's foreign policymaking and its behaviour, it seems that scholars are quite frustrated with the best fit that can be applied to such endeavours. Kim (1994) points out that it seems to be just prima facie by simply fitting or transcending almost any theoretical framework to the Chinese case, as Chinese foreign policy behaviour is so multifarious and multi-principled: new payoffs for global interaction, which in turn makes the task even more challenging.

Local Chinese scholars also seem dissatisfied with simply replicating western theories and approaches to the analysis of China's foreign policy and behaviour. Wang (1994) observed that achievements in integrating western thinking and methodology with Chinese research work have not seemed very impressive. To some extent, the splendour of western theories gets lost when one looks for their relevance to Chinese conditions, and what is worse is that Chinese scholars even see the harmful effects of applying western theories in China due to the fact that Marxism–Leninism is depreciated, devalued, and attacked, which leads to ideological hostility in their analysis (Wang, 1994).

The Micro-Macro Linkage Approach

Foreign trade policy is the result of constant interaction between international and domestic forces. In attempts to justify the policy choice, scholars have made significant efforts developing various theories, models, approaches, and frameworks, all of which have been substantiated with empirical evidence. While surely the incumbent literature has advanced the understanding of foreign trade policymaking, there is a lack of consensus as to which analytical framework is theoretically rigorous to account for both internal and external origins of trade policymaking in China (Lai, 2010).

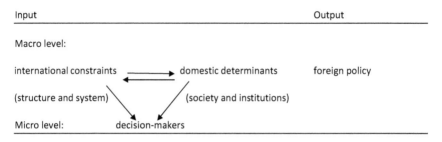

FIGURE 3.1 The Macro–Micro Linkage Approach.
Source: Zhao (1996).

It is for this reason that scholars call for the need for a more theoretically sophisticated and empirically grounded model so as to allow a comprehensive analysis embracing all elements in the policymaking process, which will aid to advance our understanding of the subject matter.

In an ambitious attempt to bridge the gap between western social science theories and the analysis of Chinese foreign policy (Hao, 1998), Zhao (1996) developed a Micro-Macro Linkage Approach (see Figure 3.1) adapted from sociological theory. He argues that the incumbent frameworks failed to identify and establish the factors at all levels, all of which have played a role in China's international policymaking. Further the incumbent frameworks either underestimated or even ignored the dynamics and the interplay of various factors, without enabling researchers to dive underneath the surface for hidden factors, thus failing to address the complexity of the policymaking process and the justification of the policy outcome.

He argues that the terms micro and macro are completely relativistic, as what is macro at one level could be micro at another.

> It is neither the macro-structure, nor the micro-decision-makers that have the absolute control over a country's foreign policy, but rather the interaction of the factors at multiple levels that should be examined in order to identify any causation, which includes international structure, domestic institutions such as national and provincial bureaucratic and social forces, as well as individual leaders.
>
> *(Zhao, 1996)*

He asserts that his formula would not only enable to examine the functions of the various factors systematically and collectively but also the correlations between them, which affect the policymaking process and the policy outcome. The advantages of aligning factors at both micro and macro levels are to avoid the shortcomings of one-sidedness, which tends to outweigh certain factors at one level to the rest at another level. By focusing on the interwoven nature of influential elements at various levels which remain active during the policy-making processes, in dynamic situations, the decisive factors can be analysed extensively including the international environment, domestic sources, and individual decision-makers, which have collectively affected the policymaking process and policy outcome. Specifically, the analysis can be undertaken to examine the following:

- The channels and mechanisms through which demands from the international and domestic environments converge on the policymaking process;
- How the evolving dynamics of domestic and international environments affect each other in shaping a country's external behaviour; and
- How they influence individual decision makers in the formulation of foreign policy.

This linkage model appreciates the various sources of influence and their coexistence at each level without discriminating one against another, with all being considered in a relative sense (Zhao 1996). He argues that none of the elements will have an absolute control over a country's foreign policy, the complexity of which can only be understood by looking into the mutual influences, channels, and mechanisms, between and among them, at both micro and macro levels.

Zhao (1996) validated his model by examining the fundamental changes in China's foreign policymaking and policy outcome from 1949 to 1995. He investigated (a) China's incremental shift from focusing on security matters to economic growth; (b) Beijing's interpretation of international and domestic environments; (c) power transition from Mao to Deng; and (d) the changes in rules, norms, and mechanisms in China's policymaking process. As a demonstration, he used Japan's official development assistance to China as a case study, with an attempt to illustrate how both countries adjusted their stance while negotiating to agree on the assistance projects.

The approach was designed to forge a new research agenda for the studies on China's foreign policy, which was radically different from the other incumbent models (Zhao, 1996). The effort is appreciated as an early attempt to bring theoretical rigour to the field of China studies (Whiting, 1994). Notwithstanding, his approach has since been little replicated or as attested through his own study, ushered a new direction for future Chinese foreign policy studies. He also developed a few agenda for future research so as to attest the framework, including "Bilateral Relations with Other Countries" and "Domestic Mood and Foreign Policy." To encourage further research, Zhao (1996) calls attention to the behaviour patterns and policy choices of Chinese foreign policy in the post-revolutionary era, which suggests that China is more likely to expand and penetrate its participation in international activities. Its pursuit for economic modernisation and regional stability would hold China to be more inclined for more extensive cooperation on security matters and economic and cultural exchanges.

Applying the Micro-Macro Linkage Approach to China's international air transport policymaking

The preceding discussion has examined the various approaches applied to the analysis of China's foreign policymaking. Although it is undeniable that the gap in interpreting the subject matter has yet been considerably narrowed (Wang, 1994) and there has always been a debate as to what is the most appropriate approach, significant achievements have been made in deepening the understanding of China's foreign policymaking behaviour, which aids to bridge the gap between the theoretical concept and actual practice.

China is unique in terms of its history, ideology, culture, bureaucracy, and social and legal systems. While it is increasingly integrated into the global economic system, it remains to be a socialist country with its own characteristics, with the Communist Party being its sole ruling party. This uniqueness not only presents the most dynamic case for study but also posits challenges when applying classical theories widely accepted by western countries to the Chinese case. A sound justification is required when theories and methodologies are determined to be appropriate to analyse the China case.

The uniqueness, though, should not be an obstacle from testing the applicability of any established theories, as Rosenau (1994) rightly argues. While it is an ongoing debate as to which will be the best fit, scholars consent that caution is essential when exercising the reapplication in attempts to interpret its behaviour and recurrent patterns. Shambaugh (1994) calls for a need "for Western sinologists to escape from their ethnocentrism, instead, to attempt to crawl inside the mindset of the Chinese elites and ordinary people, which should be achieved at no cost of scholarly objectivity or universal standards of human rights." Hamrin (1994) calls for the Sinologists and researchers interest in China to "find ways to integrate studies on China's behaviour on domestic and international matters adopting a longitudinal and interdisciplinary approach." The various sources, though likely competing for the attention of the decision makers, are all worth studying. These include ideological preferences, perceptions of China's national interest, assessment of China's material power, historical experience, political tradition and political culture, domestic social structure, and political attributes, such as the evolving political climate, and the composition of the leadership (Hsiung, 1980). Wang (1994) suggests a Sinocentric approach by calling for scholars to integrate the study on Chinese politics into the domain of comparative political studies, which would enrich the subject matter. Although admittedly no theoretical framework available provides a satisfactory and less conclusive answer to the questions as to which one is the best fit, subject to individual analytical needs (Whiting, 1994), studies on China need to be contextualised (Rosenau, 1994).

Although Zhao's approach is developed to analyse China's foreign policymaking, it is considered appropriate for it to be applied to analyse China's international air transport policymaking. The rationale for such an application is based on the following justifications:

- International air transport has long been considered as a political tool in China to support the country's diplomatic and political objectives. Before the 1970s, China had very few international routes, which only connected the country to its socialist allies. International air transport negotiation was used as a precursor to ascertain the stance of the foreign government to recognise One China policy. Bilateral Air Service Agreement was regarded as one of the diplomatic instruments that facilitated the establishment of the official diplomatic ties with China. Such a political feature only started to fade in the 21st century after China joined the WTO, with the international air transport being categorised as service industry, and the government committing itself to removing trade barriers to promote the liberalisation of free trade.
- International air transport policy, even today, still remains within the domain of China's overall foreign policy realm, with the responsible authority reporting to the State Council and MoF for policy guidance and the principles governing the bilateral negotiations. Considerations are primarily given to whether the air transport policy towards a specific nation state is in alignment with its overall foreign policy/diplomatic objectives and the relationship with the country.
- A third justification is that international air transport is governed by international regime, namely the Chicago Convention. The ICAO sets the rules and regulations that China has to abide by as a signatory member state.

References

Abe A., (1999), *Japan and the European Union, domestic politics and transnational relations*, The Athlone Press, London.

Baldwin R. E., (1996), The political economy of trade policy: integrating the perspectives of economists and political scientists, in: Feenstra R. C., Grossman G. M., and Irwin D. A. (editors), *The political economy of trade policy: paper in honour of Jagdish Bhagwati*, (pp. 147–174), The MIT Press, Cambridge, MA.

Bayne N., and Woolcock S., (2007), What is Economic Diplomacy, in: Bayne N., and Woolcock S., (editors), *The new economic diplomacy, decision-making and negotiation in international economic relations*, (2nd edition), (pp. 1–20), Ashgate Publishing Limited, Aldershot.

Cohen B. J., (1968), (editor), *American foreign economic policy: essays and comments*, Harper and Row, New York.

Cohen S. D., (2000), *The making of United States international economic policy: principles, problems, and proposals for reform*, Praeger Publishers, Westport, CT.

Dam K. W., (2001), *The rules of the global game, a new look at us international economic policymaking*, The University of Chicago Press, Chicago, IL.

Dumbrell J., (1997), *The making of US foreign policy*, (2nd edition), Manchester University Press, Manchester.

Edwards G. C., and Wayne S. J., (2009), *Presidential leadership: politics and policy making, Wadsworth Cengage Learning*, Wadsworth Cengage Learning, Boston, MA.

Frieman W., (1994), International science and technology and Chinese foreign policy, in: Robinson T. W., and Shambaugh D., (editors), *Chinese foreign policy, theory and practice*, (pp.158–198), Oxford University Press Inc., New York..

Gilpin R., with the assistance of Jean M. Gilpin, (2001), *Global political economy: understanding the international economic order*, Princeton University Press, Princeton, NJ.

Goldstein J., (1988), Ideas, institutions, and American trade policy, *International organisation*, 42 (1): pp.179–217.

Hamrin C. L., (1994), Elite politics and the development of China's foreign relations, in: Robinson T. W., and Shambaugh D., (editors), *Chinese foreign policy, theory and practice*, (pp. 70–114), Oxford University Press Inc., New York.

Hao Y., (1998), Interpreting Chinese Foreign Policy: The Micro-Macro Linkage Approach, (book review), *American Political Science Review*, 92 (2), June, pp. 510–511.

Hao Y., and Su L., (2005), (editors), *China's foreign policy making, societal force and Chinese American policy*, Ashgate Publishing Limited, Aldershot.

Hsiung J. C., (1980), The study of Chinese foreign policy: an essay on methodology, in: Hsiung J. C., and Lim S. S., (editors), *China in the global community*, (pp. 4–6), Praeger, New York.

Ikenberry G. J., Lake D. A., and Mastanduno M., (1988), *The state and American foreign economic policy*, Cornell University Press, Ithaca, NY.

Johnson U. A., (1972), *National security policy and the changing world power alignment, hearing-symposium developments*, 92nd Congress, 2nd Session, May–August, 1972, p. 368, US Congress, House, Committee on Foreign Affairs.

Kerr W. A., (2007), Introduction to trade policy, in: Kerr W. A., and Gaisford J. D., (editors), *Handbook on international trade policy*, Edward Elgar Publishing Limited, Cheltenham.

Kim S., (1994), China's international organisational behaviour, in: Robinson T. W., and Shambaugh D., (editors), *Chinese foreign policy, theory and practice*, (p. 405), Oxford University Press Inc., New York.

Lai, H., (2010), *The domestic sources of China's foreign policy: regimes, leadership, priorities and process*, Routledge, Oxon.

Lampton D. M., (2001), *Same bed, different dreams: managing US-China relations, 1989–2000*, University of California Press, Berkeley and Los Angeles.

Levine S. I., (1994), Perception and ideology in Chinese foreign policy, in: Robinson T. W. and Shambaugh D., (editors), *Chinese foreign policy, theory and practice*, (pp. 30–46), Oxford University Press Inc., New York.

Lipson C., and Cohen B. J., (1999), *Theory and structure in international policy economy, an international organisation reader*, The MIT Press, Cambridge, MA.

Muscatelli V. A., (1996), *Economic and political institutions in economic policy*, Manchester University Press, Manchester and New York.

Ng-Quinn M., (2004), The analytic study of Chinese foreign policy, in: Liu G., (edited) *Chinese foreign policy in transition*, (pp. 32–60), Transaction Publishers, New York.

Odell J. S., (1990), Understanding international trade policies, an emerging synthesis. *World Politics*, 43, 140–157.

Pastor R. A., (1980), *Congress and the politics of U.S. foreign economic policy 1929–1976*, University of California Press, Berkeley and Los Angeles.

Persson T., and Tabelline G., (2000), *Political economics, explaining economic policy*, The MIT Press, Massachusetts Institute of Technology, Cambridge, MA.

Rosenau J. N., (1994), China in a bifurcated world: competing theoretical perspectives, in: Robinson T. W. and Shambaugh D., (editors), *Chinese foreign policy, theory and practice*, (pp. 524–554), Oxford University Press Inc., New York.

Shambaugh D., (1994), Patterns of interaction in Sino-American relations, in: Robinson T. W. and Shambaugh D., (editors), *Chinese foreign policy, theory and practice*, (pp. 197–223), Oxford University Press Inc., New York.

Stein A. A., (1990), *Why nations cooperate: circumstance and choice in international relations*, Cornell University Press, New York.

Suettinger R. L., (2003), *Beyond Tiananmen, the politics of US-China relations, 1989–2000*, Brookings Institution Press, Washington, DC.

Tow W. T., (1994), China and the international strategic system, in: Robinson T. W. and Shambaugh D., (editors), *Chinese foreign policy, theory and practice*, (pp. 115–157), Oxford University Press Inc., New York.

Wang J., (1994), International relations theory and the study of Chinese foreign policy: A Chinese perspective, in: Robinson T. W. and Shambaugh D., (editors), *Chinese foreign policy, theory and practice*, (pp. 481–505), Oxford University Press Inc., New York.

Whiting A.S., (1994), Forecasting Chinese foreign policy: IR Theory vs. the fortune cookie, in: Robinson T. W. and Shambaugh D., (editors), *Chinese foreign policy, theory and practice*, (pp. 526–523), Oxford University Press Inc., New York.

Woolcock S., (2007), Theoretical analysis of economic diplomacy, in: Bayne N., and Woolcock S. (editors), *The new economy diplomacy, decision-making and negotiation in international economic relations* (2nd edition) (pp. 21–42), Ashgate Publishing Limited, Aldershot.

Zampetti A. B., (2006), *Fairness in the world economy, US perspective on international trade relations*, Edward Elgar Publishing Limited, Cheltenham.

Zhao Q., (1996), *Interpreting Chinese foreign policy, the micro-macro linkage approach*, Oxford University Press (China) Ltd., Hong Kong.

Zweig D., (2002), *Internationalising China, domestic interests and global linkages*, Cornell University Press, Ithaca.

4

CHINA AND THE USA

The USA's international air transport industry

Its vast landscape and robust commercial environment has kept sustained air travel demand in America, which boasts the world's largest single air transport market, with around two million people, 50,000 tons of cargo, and more than one million bags travelling onboard 25,000 flights every day (ATA, 2010). According to the DoT, around 750 million inbound and outbound scheduled passengers travelled between the USA and other countries in 2004 compared to 965 million in 2017, representing an almost 28 percent increase (DoT, 2018). The majority of trips in the USA are domestic travel which accounted for 76 percent of passenger traffic in 2017.

Air transport remains of significant importance to the US economy, generating more than $530 billion in gross value added contribution to GDP. It created almost $2.2 million direct jobs and a further $1.2 million indirect jobs, and contributed to $2.4 trillion in trade value by air freight (Oxford Economics, 2016). In 2008, the US airlines carried about 4.4 million tons of international air cargo, accounting for 18 percent of the 25 million tons transported globally in international service. Its international traffic generated 33 billion RTK, accounting for 25 percent of about 131 billion RTK of the global international air cargo traffic (DoT, 2009). In 2009, the industry recorded 769.5 billion revenue passenger miles (RPM) and 25 billion revenue tonne miles (RTM) with 703.9 million passengers enplaned (ATA, 2010). A study commissioned by the Federal Aviation Administration (FAA) in 2016 found that the industry's economic impact stood at $1.6 trillion and supported approximately 11 million jobs (directly and indirectly) during 2012–2014.

Developing out of the legacy of the two world wars, US airlines have sustained their positions as globally competitive entities. Through mergers and alliances, they have dominated various leagues of the world's top businesses. Of the Fortune 500 businesses in 2018, 8 were US carriers, with United Parcel Services (UPS) taking 44th place and FedEx 50th. Of the eight US carriers, these two cargo carriers preceded the remaining six passenger airlines, namely American Airlines (71st), Delta (75th), United Continental Holdings (81st), Southwest Airlines (142nd), Alaska Air Group (355th), and Jetblue Airways (402nd). Of the world top ten airlines in 2017 measured by operating revenue, four were US carriers including Delta Airlines (DL) (hereinafter referred to as Delta), American Airlines Group, United Continental Holdings, and Southwest Airlines. According to the IATA World Air Transport Statistics, when measured by RPK, four out of the top ten airlines were also US carriers with American Airlines taking the lead reporting 323,968 RPKs followed by Delta and United Airlines. Chinese carriers China Southern, China Eastern, and Air China were ranked sixth, ninth, and tenth, respectively.

As the cradle of the modern aviation industry, the US airlines have remained at the forefront of competition by championing innovative strategies and marketing initiatives. For instance, the low-cost carrier business model (LCC) first appeared in the USA after the industry's deregulation in 1978, which has since revolutionised the way people perceive air travel. The hub-and-spoke system was initially adopted by US carriers in order to optimise their resources to achieve economies of scale. They were the first to develop a frequent flyers programme (FFP) to retain the loyalty of passengers to secure traffic. They were the first to develop the computer reservation systems (CRS) so as to allow airlines to monitor their seat booking, control the inventory, and distribute the sales of seats among travel and sales agents to optimise yield. They were the first to formulate codeshare arrangements and alliances to bypass regulatory constraints and circumvent the restrictions of foreign investment in national airlines to achieve international expansion.

Though initially resisting deregulation in the late 1970s and early 1980s, the industry has since enjoyed the benefits of operating in a non-restrictive environment, which has enabled them to expand rapidly into international markets. They reversed their attitude towards economic regulations and became the biggest advocate of liberalisation by lobbying the government for as many open skies deals with the rest of the world as possible, so as to allow them to take advantage of penetrating into these markets for global coverage.

The majority of the large carriers such as United, Continental (part of United since 2010), Delta, Northwest (part of Delta since 2007), American, FedEx, UPS, and Polar Air Cargo have flown to China. United bought its China route licence from Pam Am in 1986 and remained the biggest carrier in the Sino-US market until 2017 in terms of capacity provided. In 2018, Air China surpassed United as the largest carrier in the China-US market by 68,540 seats (See Figure 4.1 FedEx and UPS operate all-cargo

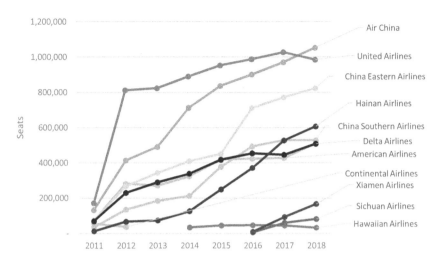

FIGURE 4.1 2011–2018 US–China capacity.
Source: www.caac.gov.cn; Centre for Aviation Database.

flights between the two countries with their regional hubs being established in Guangzhou and Shanghai in 2005 and 2007, respectively, which enable them to further expand their freight operation in the Asia-Pacific region. Table 4.1 summarises the US carriers' operation in the Sino-US markets up to 2014.

The USA also lays claim to the busiest airports of the world. In the domestic market, Atlanta Airport was the busiest airport in the USA in 2018 with 51.4 million passengers enplaned, followed by Los Angeles with 42.5 million and Chicago's O'Hare with 39.8 million (Bureau of Transportation,2018). The most recent data available from ACI for 2017 highlights that of the world top ten airports measured by passenger throughputs in 2017, 3 were US airports down from five a decade ago. This change has largely been precipitated by the rising prominence of Asian and Middle Eastern airports. In 2009 only two of the top ten airports by passenger throughput were from Asia (namely Tokyo Haneda and Beijing), five from the USA (Atlanta, Chicago O'Hare, Los Angeles, Dallas Forth Worth, and Denver), and the remaining three from Europe (Heathrow, Paris, and Frankfurt). In 2018, the top ten airports by passenger throughput comprised three US airports (Atlanta, Chicago O'Hare, and Los Angeles), four Asian (Beijing, Shanghai, Hong Kong, and Tokyo Haneda), one Middle Eastern (Dubai), and two European airports (Paris and Heathrow) (Figure 4.2). In terms of cargo traffic (see Figure 4.3), Memphis ranked world No. 1 recording 3.7 million tonnes of freight in 2009, with Anchorage taking the sixth place with two million tonnes, and Louisville the seventh with 1.95 million. By 2017, however,

TABLE 4.1 US designated carriers in the Sino-US markets by the end of 2014

Carrier	Year into market	Destinations in China
United	1986	Beijing
		Shanghai
		Guangzhou
Northwest	1984	Beijing
		Shanghai
		Guangzhou
Continental	2006	Shanghai
		Beijing
Delta	2008	Shanghai
American	2006	Beijing
FedEx	1993	Shanghai
		Beijing
		Shenzhen
		Guangzhou (hub)
UPS	2001	Shanghai (hub)
		Guangzhou
Polar Air Cargo	2004	Shanghai
		Beijing
Evergreen International	2009	Shanghai
Kalitta Air	2010	Shanghai
Hawaiian Airlines	2014	Beijing

Source: www.caac.gov.cn; Centre for Aviation Database.

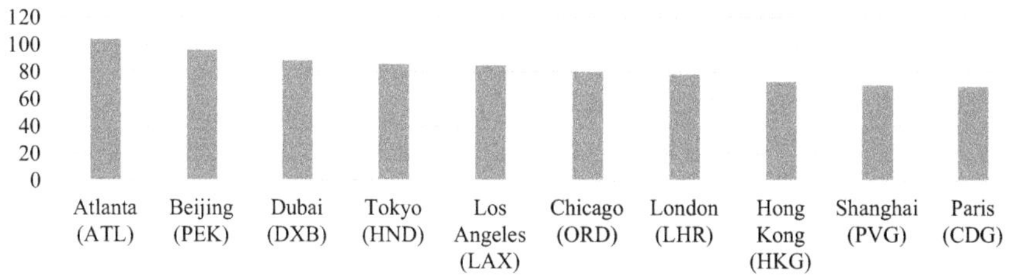

FIGURE 4.2 World's top 10 airports measured by passenger throughputs in 2018 (in millions).

Source: ACI.

data by ACI reveals that Memphis' dominance was replaced by Hong Kong, which facilitated 700,000 tonnes of cargo more than Memphis. Of the total of more than 500 airports in the USA serving commercial flights, 12 had direct services to China which include New York, Chicago, Detroit, Atlanta, Seattle, Los Angeles, San Francisco, and Honolulu for both passengers and cargo while

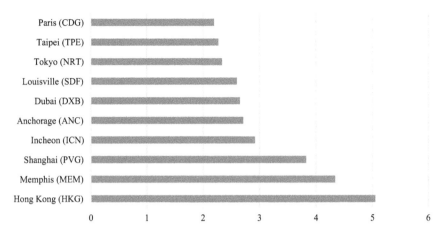

FIGURE 4.3 World's top 10 airports measured by cargo volume in 2018 (in million tonnes).

Source: ACI.

Memphis, Portland, Anchorage, Chicago, Orlando, New York, Los Angeles, and San Francisco for cargo operations.

Champion in liberalising its domestic market and in promoting open skies in the international air transport market

The USA is undoubtedly the world champion in liberalising both its domestic and international air transport markets and has been persistently pressing other countries to follow suit. Back in 1944 when the Chicago Conference was held, the USA was one of the few nations that advocated a free regime for international air transport to allow competition for efficiency and productivity. It argued that a multilateral mechanism towards international air transport operations would nurture efficiency in the industry which would benefit the global economy. As an economic system driven by capitalism and the free movement of wealth, goods and services, liberalisation of the airline industry would also undoubtedly help sustain prosperity and the security of the US economy. However, due to the opposition from other nations that had concerns about being taken advantage of by US carriers, the USA failed to convince the attendees at the Chicago Conference to accept its proposals. Two years later, it concluded the Bermuda I agreement with the UK, which set the model of bilateral negotiations on international air transport operation, whereby a nation state's government would act as the key negotiator and decision maker to determine how airlines should operate cross-border flights.

The USA has never given up its efforts to remove the restrictions on airline operations. Inspired by the liberal ideas advocated by politicians and economists in the 1970s, the US government took the initiative to demolish the economic regulatory controls on the operations of its airlines, thus being the first to have deregulated its domestic air transport market. With the determination of Presidents Carter and Ford, the support of senators, bureaucrats in the Civil Aviation Board (CAB), and academics, the USA passed the "Airline Deregulation Act" in 1978, which removed the government's interference in airline operations to allow free competition, in particular with respect to route entry, capacity, and pricing, despite the move being harshly resisted by airlines, airports, and other stakeholders.

The empirical evidence of the benefits of liberalising the domestic market has convinced the US government to pursue a more liberalised approach to international market. It officially began pivoting its international air transport policy towards contemporary open skies in 1995. The policy was designed to knock down the protection barriers to allow the US carriers to operate in an open and free environment globally. The US government envisioned an environment where it would liberalise international transportation markets, extend passenger and freight capacity, and support the growth of technology and infrastructure. It portrayed a quest towards liberalisation that would be expected of a nation with an economic system built on the premises of free trade.

Efforts to reduce barriers to trade in air transport have remained paramount for the US government to maintain its leadership by enhancing the competitiveness of US transport providers and manufacturers globally. Indeed, ensuring the global competitiveness of the US' aviation industry is central to the guiding principle of the US DoT as epitomised through its mission statement which is to:

> Ensure our nation has the safest, most efficient and modern transportation system in the world; that improves the quality of life for all American people and communities, from rural to urban, and increases the productivity and competitiveness of American workers and businesses.
>
> *(US DOT 2020)*

Liberalisation efforts have undoubtedly brought about substantial rewards. In just less than 30 years after the signing of the first such open skies agreement with the Netherlands in 1992, the USA has negotiated more than 120 bilateral open skies agreements with partner nations in all geographical regions. These agreements span from the smallest of nations, some of which have no domiciled carriers such as St. Vincent and the Grenadines, to more consequential air transport partners such as Canada, Japan, and Australia.

One of those countries that have not concluded such agreements is China. With a population of well over 1.4 billion and a GDP of $12 trillion US dollars by 2017 acccording to statistics in the Worldbank's databank), China's passenger aviation market has been forecasted by IATA to be in a position to replace the USA as the world's largest aviation market by as early as 2024. China's growth creates significant opportunities for US airlines not only for its O–D (origin–destination) demand potential, but also its positioning in the Asia-Pacific region and potential for beyond traffic through fifth freedom rights. This is particularly significant since in addition to China surpassing the USA as the largest traffic market, IATA also estimates that the Asia-Pacific region will be the biggest driver of air transport demand towards 2035, accounting for in excess of half of all air transport traffic.

Over the years, the bilateral trade between China and the USA has continuously grown, with China overtaking Japan to become the second largest trade partner of the USA and holding a tremendous amount of its foreign debts. The booming of the bilateral trade has benefited the air transport industry, with the US carriers diverting more of their focus to this market and the Asia-Pacific region, which passed North America as the world's largest regional aviation market in 2009 and would remain to be the most profitable region for the world's airlines (IATA, 2010). The market's significance cannot be understated; indeed, some authors such as Steinberg (2007) have suggested that it represents all the potential of the growth of the industry for the US carriers. To this effect, it is perceived that the USA would like to see an open skies agreement to be achieved within a foreseeable timescale with China so as to enhance the overall comprehensive cooperation between the two parties.

The US government agencies responsible for international air transport policymaking and bilateral negotiations

Due to the nature of the primacy of international law over domestic law in aviation and the international aspects of air politics and diplomacy, different countries have developed different regimes and structures concerning international aviation matters. The USA sees air transport negotiations bear political as well as economic implications (Havel, 2009), hence the Department of State (DoS) has the overall responsibilities by taking a leading role in all ASA negotiations, as supported by the DoT and the Federal Aviation Administration (FAA). The president of the USA has the final say on US policy decisions on international aviation matters.

There are two offices within the Bureau of Economic, Energy and Business Affairs (EEB/TRA) in DoS that are specifically responsible for international aviation matters, which are the Office of Aviation Negotiations (OAN) and the Office of Transportation Policy (OTP), which work together to drive US air transport policy initiatives. Headed by the chief negotiator, the Office of Aviation Negotiations (OAN) manages US bilateral aviation relationships and leads the open skies and other aviation liberalisation agreements negotiations with its counterparts. The Office of Transportation Policy (OTP), develops and coordinates policies on international civil aviation, maritime, and land transport issues, including policy research, safety and security, discriminatory and unfair practices, commercial, and operational problems encountered abroad, overflight/

landing authorisations, port access, environmental protection, and accident investigations. The OTP also participates in the development of policies concerning international civil aviation matters with substantive and multinational applications, such as those involving ICAO and the EU.

The DoT is the industry regulator. Through the Office of Aviation & International Affairs, under the Secretary of Transportation, the DoT provides the following services: providing policy advice and policy recommendations to the assistant secretary for aviation on issues involving international civil aviation, including bilateral and multilateral aviation negotiations; serving as the department's representative at formal and informal bilateral and multilateral air transportation negotiations and providing policy guidance and technical and analytical support to other US agencies during negotiations.

The FAA oversees the safety aspect of civil aviation. The Office of Aviation Policy, Planning and Environment leads the agency's strategic policy and planning efforts, coordinates the agency's reauthorisation before Congress, and is responsible for national aviation policies and strategies in the environment and energy arenas, including aviation activity forecasts and economic analysis. The Office of Government and Industry Affairs is the principal advisor and representative on matters concerning the Congress, aviation industry groups, and other government organisations. The FAA is also involved in bilateral aviation negotiations, though only playing a supporting role.

China–USA trade relations

China has emerged as a leading global manufacturer and a driving force for international trade. According to World Bank data, the past few decades have seen its trade volume increasing exponentially with its merchandise exports totalling USD 2.26 trillion in 2017, compared with USD 1.2 trillion in 2009 while the imports of goods and services achieved USD 1.03 trillion and USD 2.2 trillion, respectively. In 1989, only 24.4 percent of Chinese exports went to the USA, but the figure jumped to 38.7 percent in 1998, resulting in China becoming the biggest trading partner of the USA. Since then, China has remained the top trade partner for the USA with export values reaching $430 billion in 2017, accounting for 19 percent of its overall exports (Figure 4.4).

Notwithstanding the scale of the increase in export value, it is worth nothing that percentage-wise, China's exports to the USA, relative to its remaining trading partners has remained stable, which underscores China's overall growth as a manufacturing economy and its significance in global trade. From the USA's perspective, trade with China of $18 billion in 1989 had grown to $310 billion in 2009, positioning China as the USA's third largest export market, accounting for 6.5 percent of US exports behind Canada (19 percent of US exports) and Mexico (12 percent of US exports). As of 2017, it remained the third largest export partner accounting for approximately 130 Billion (8 percent) of all exports.

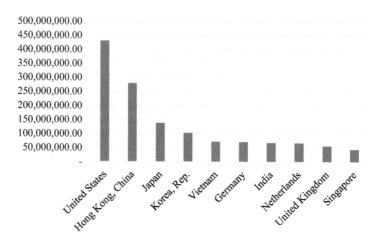

FIGURE 4.4 US's Top 10 trade partners in 2017.
Source: www.data.worldbank.org.

Between 1997 and 2017, US imports from China increased from $65 billion US dollars to $526 billion, a growth of 696 percent. This is contrasted with a 900 percent increase in US exports to China. China's accession to the WTO in 2001 allowed the country to reduce barriers to the importation of products from other nations, leading to a marked increase in China–USA trade. For instance, during the period of 2001 to 2006, the years succeeding the immediate accession, China's trade rose by 23 percent on average per annum (Beningo, 2008). This is contrasted with an average annual growth rate of 14 percent per annum for the five-year period preceding the accession. The increase was phenomenal when compared with the US trade with other East Asian countries such as Japan and South Korea, which on average grew at only 3 percent per annum. Figure 4.5 shows Sino–US trade between 1997 and 2017 and the extent to which US imports exceeded its exports to China.

China's trade expansion was not only contributed to its rapid continuous economic growth, but also by the type of manufactured products the country has invested in over the years (Tanger, 2007). The sophisticated electronic and high-tech goods including computers, mobile phones, and flat-screen televisions that are air shipped represented 58 percent of China's exports to North America and Europe compared to a 38 percent share in 1995 (Tanger, 2007). On the other hand, imports from the USA to China typically include intermediate materials and capital equipment that support China's manufacturing requirements, thus leading to an imbalance in bilateral trade. As a consequence, the trade deficit grew rapidly from minus $83.7 billion in 2000 to $240 billion in 2009 for the US side, widening further to $396 billion by 2017. Figure 4.5 expressing the trade values and the changes between the two nations from 1997 to 2017 also reveals the dramatic increase of the US trade deficit with China over the period.

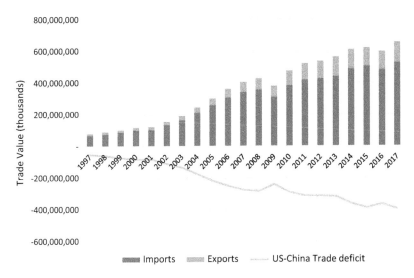

FIGURE 4.5 US imports from and exports to China between 1997 and 2017.
Source: www.data.worldbank.org.

China's outbound tourism to the USA

Compared with the trade of merchandised goods, the movement of people between the two countries has not been as impressive. Although tourism started in China in the early 1950s, the country had remained an inbound destination until the 1980s as outbound travel by Chinese citizens for leisure purposes was not permitted (Lim and Wang, 2008). One of the reasons is that the movement of people is not merely an item of trade but also involves issues of citizenship, visas, and migration (Clancy, 2007). In the late 1980s, around 95 percent of airline passengers in China were foreigners with only 5 percent being Chinese citizens (Lampton, 2001). Compared with South Korea which has had a more balanced inbound and outbound traffic, between 1996 and 2000, China had consecutively had almost four times more inbound visitors than its outbound tourists (Zhang and Chen, 2003), (Table 4.2).

The situation started to change since the late 1990s with China's marked economic growth which has improved the living standards of Chinese people and generated a burgeoning middle-class group, thus stimulating the demand for air travel. The gradual relaxation of regulatory restrictions on international air transport has enabled the growth of China's outbound travel market, with the World Tourism Organisation estimating that China would be the world's fourth largest international tourism market by 2020, representing 6.4 percent of the total, with nearly 100 million outbound travellers (Lim and Wang, 2008). Data

TABLE 4.2 International inbound/outbound tourists in three nations in Northeast Asia (in thousands)

Year	China		Japan		South Korea	
	Outbound	*Inbound*	*Outbound*	*Inbound*	*Outbound*	*Inbound*
1996	5,601	22,765	16,694	4,244	4,649	3,684
1997	5,936	23,770	16,802	4,669	4,542	3,908
1998	6,505	25,073	15,806	4,556	3,067	4,250
1999	9,200	27,047	16,357	4,901	4,342	4,660
2000	10,500	N/A	17,818	5,272	5,508	5,322

Source: Adapted from Zhang and Chen (2003).

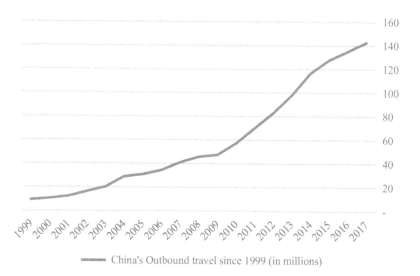

China's Outbound travel since 1999 (in millions)

FIGURE 4.6 China's outbound travel since 1999 (in millions).
Source: stats.gov.cn.

from the CNTA[1] showed that between 1995 and 2008, the number of Chinese outbound travellers increased from 4.5 million to 45.8 million, averaging more than 10 percent increase year on year (Figure 4.6).

The USA has always been one of the most popular destinations for Chinese outbound travellers although the Approved Destination Status (ADS) had not been granted until December 2007. The ADS not only recognises the USA as a tourist destination for Chinese leisure travellers but also endorses the role of tour operators who could now officially market and promote US destinations in China and organise group tours for Chinese leisure travellers.

In 2007, around 397,000 Chinese outbound travellers visited the US, compared to 199,000 in 1996, making the China-US country pair the only long-haul market in the top ten Chinese outbound tourism destinations. Total visitor spend was $2.7 billion (US Department of Commerce, 2007). Of the 397,000 visitors that year, 58 percent were leisure travellers, despite the fact that there was no visa-waiver system between the two countries and restrictions on the sale of leisure travel were still in place (US Department of Commerce, 2017). A decade later in 2017, 3.2 million Chinese visitors visited the USA with expenditure totalling $35.3 billion with approximately 58 percent representing leisure travel (US Department of Commerce, 2017).

The US outbound tourism to China

Statistics from the National Tourism and Travel office shows that US visitors to China have been growing steadily over the years, making China one of the most sought-after destinations for US citizens. In 1996, the total number of US outbound visitors to China stood at 396,000, which increased to 1.37 million in 2007, making China the 16th largest international tourist destination market for the USA. From China's perspective, the USA was the fourth largest tourist source market, just behind South Korea, Japan, and Russia. The following five years saw a continuous increase of Chinese outbound visitors to the USA, which exceeded the US outbound visitors to China in 2012 (see Figure 4.7). By 2017, there were approximately 3.2 million Chinese trips to the USA compared to the 1.3 million from the USA to China. Between 2010 and 2011, there was an increase in Chinese visitation of 35 percent compared to a 12 percent decline from the US side. US arrivals have thus consistently averaged around 1.2 million, showing only slow growth compared with the growth witnessed in the market of visitors from China. The substantial rise in Chinese visitation,

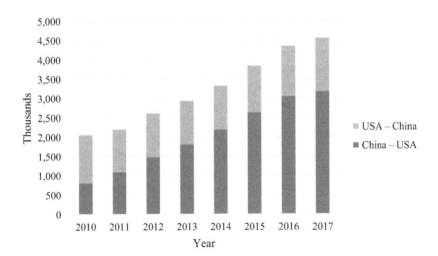

FIGURE 4.7 Visitors between China and USA, 2010–2017.

Source: travel.trade.gov.

while commensurate with an overall increase in Chinese outbound tourism was also influenced by relaxations of travel restrictions on Chinese citizens. A revised visa agreement reached between the USA and China extended visa validity from one to ten years.

Latest trade development between the two countries

The expansion of economic ties, in particular bilateral trade, brought with it both positive and negative consequences for both countries and their relations (Lampton, 2001). Positively, the USA's increased direct investment in China, which stood at $82.7 billion in 2007, helped create a significant number of jobs in the USA, while made-in-China products provided US consumers with low-cost, good-quality products. The US multinational corporations such as Boeing, Coca Cola, and Motorola are able to develop products and services for international consumers and expand into the local Chinese market with a growing market share. It also helped China to develop its export-oriented economy, accumulate foreign reserves, and facilitate its collaboration with the international community. Negatively, the economic interdependence, in particular, the trade imbalance, develops into trade friction and carries with it political implications, which have increasingly become pressing issues challenging both parties. Demanding market access for US enterprises is one of the themes that the US side would always request in communications with the Chinese side, with the aims of generating more opportunities for the US enterprises to penetrate in Chinese market to reduce the trade imbalance.

However, as has been attested to herein, aviation is not insulated from the geopolitical arena, and tensions in trade relations between China and the USA even further accentuated these challenges. During the lead-up to the 2016 presidential elections, then Presidential candidate Donald Trump espoused a foreign policy stance that was increasingly critical of China. After being elected to the Presidency, the Trump Administration and China began negotiations in attempts to resolve trade dispute which concluded with a trade deal in May 2017 that expanded markets in some respects to both nations. Almost a year later, in March 2018, President Trump signed a memorandum which, inter alia, imposed tariffs on Chinese products. China retaliated in April 2018 proposing its own round of tariffs on some US-made products. A brief truce to the ongoing trade war was established following agreements by China to increase purchases of US goods but subsequently reinstated after talks were said to have broken down. Between mid-2018 and 2019, the "tit-for-tat" continued between the two nations. Indeed, the very existence of some facet of "tit-for-tat" is essential to validating the presence of trade wars (Zeng, 2004). Notwithstanding talks seemingly progressing, and more extensive truces being agreed in between as negotiations continued, by May 2019, President Trump threatened to once again add further tariffs after declaring that China had attempted to renegotiate and renege on agreements made towards the trade deal.

Enhancements to trade relations between the two nations have had immediate implications for fostering greater ties between the respective aviation industries. Some efforts to extend trade relations in aviation precede the current trade war, when, in 2015, the US manufacturer Boeing and state-owned Chinese manufacturer COMAC entered into a collaborative arrangement to develop a 737 "completion centre" in Beijing. Additionally, though the deal had been in development for some time, in 2017, during a visit to Beijing as a part of the first round of trade negotiations, President Trump touted China's agreement to purchase 300 Boeing aircraft at list prices of US$ 37 billion as testimony of his capacity to narrow the gap in the USA–China trade deficit. The deal was signed between state-owned China Aviation Supplies Holding Company, which leases aircraft to Chinese carriers, and American manufacturer Boeing. On 25 March 2019, China Aviation Supplies Holding Company entered a similar agreement with European manufacturer Airbus to purchase a similar quantum of aircraft with list prices of US $2 billion, less than the deal with Boeing at US $35 billion. Carrying a virtual duopoly in large aircraft manufacturing, the competition between Boeing and Airbus is naturally intense. As China–USA trade tensions continue to evolve, the consequential impact on trade relations for the aviation industry cannot be understated. Results of May 2019 negotiations between the two governments in attempts to reach consensus on a trade deal brought this reality into sharp focus. On 10 May 2019, after negotiation talks stalled, the US government proceeded with plans to ascribe tariffs ranging from 5 to 25 percent on a further USD $200 billion of imports from China.

The additional tariffs, due on imports from 1 June 2019 reinforced perceptions by some economists that although the trade war would negatively impact economic growth within the US economy, China would likely suffer more significantly. This is in large part attributable to China's trade surplus with the USA. In response to the May 10th tariff increases, China levied its own supplemental round of tariffs, also implementable from 1 June 2019, on approximately 5,000 US products equating to Chinese imports of USD $60 billion. Significantly however, it was reported that in addition to this planned increase, China was considering reducing its orders of Boeing aircraft. Coupled with safety concerns with the 737 Max aircraft, the aircraft manufacturer was now positioned to become embroiled within the ongoing trade dispute.

China–USA air transport relations

Since 1980, when the China–USA bilateral ASA was signed, six major amendments have been made in 1982, 1992, 1995, 1999, 2004, and 2007, respectively, which have substantially relaxed the restrictions on airline operations. Barriers to market entry were removed, capacity was increased, and price control was lifted which has enabled unlimited designation of carriers operating between the two countries to satisfy the market demand. The following is a detailed examination of the bilateral air transport relations which reveal how the restrictions have been gradually eliminated to embrace a liberalised trade deal between the two countries.

Establishment of bilateral air services relations

There were no diplomatic ties between China and the USA between 1949, when the new PRC was established, and 1972, when the then American President Richard Nixon extended his hand to Premier Zhou En'lai, indicating a willingness to end the hostility of more than 20 years between the two countries. However, Nixon's historical visit only brought about the first order of ten Boeing 707 aircraft from the Chinese side to serve its market rather than the rapprochement of diplomatic ties due to the divergent and contrasting views on certain issues. It took another six years for both parties to conclude the Joint Communique of China and the USA concerning the establishment of their diplomatic relations in December 1978, which declared that "Starting from January 01, 1979, the Chinese and American sides will acknowledge each other and establish diplomatic relations," thus putting an end to almost 30 years of mutual estrangement between the two countries.

In the 1980s, China started its economic reform and was very keen to attract foreign direct investment, to import the technology, and to promote its exports to the western countries including the USA. As an element of the diplomatic relationship, a bilateral ASA was signed in Washington D.C. in September 1980 by the visiting Chinese Vice Premier Bo Yibo and the then US President Jimmy Carter. Unlike most of the ASAs in those days that China had concluded which only permitted one flag carrier to fly between the capital cities, the deal with the USA allowed two carriers from each side to operate on two routes (Route A and Route B) (Table 4.3) with two frequencies per week. For the US side, both the capital city Beijing and the most commercial city of Shanghai were allowed for operations, while for the Chinese side, four cosmopolitan cities rather than the capital city of Washington were given, namely New York, Los Angeles, San Francisco, and Honolulu. Although Route B was

TABLE 4.3 Sino-US arrangements in 1980

Items	*Agreement*
Route A for both sides	Beijing–Shanghai–Tokyo/or another point in Japan–Honolulu–Los Angeles–San Francisco-New York, And v.v.
Route B for both sides	Beijing–Guangzhou–Shanghai–Tokyo/or another point in Japan–Honolulu or Seattle–Los Angeles–San Francisco–Chicago, And v.v.
Number of designated carriers	2 for each side
Frequencies	2 weekly frequencies for each carrier

Source: Adapted from Sino-US ASAs in 1980 and 1982.

in principle agreed in the 1980 discussions, the specific points enroute were not agreed until 1982 when a second consultation was held with Guangzhou being included to the route.

The double designation arrangements were exceptional in the early 1980s from the Chinese perspective. On the one hand, China's air transport system was regarded as quasi-military services. There were no airlines existing in China that were managed as a corporate business to explore the commercial benefits of air transport. The CAAC, as a subordinate organisation reporting to the Air Force, acted as a regulator as well as a commercial operator with the IATA code CA. On the other hand, air transport in China was considered as more a political instrument rather than an economic activity. International air links were only made available to those countries that had diplomatic ties with China, with Chinese people travelling by air being strictly scrutinised.

In such a circumstance, double designations for the Sino-US market were not meaningful for the Chinese side as it was for the US, since there was no second carrier available at all in the market to be designated by the Chinese authority. The CAAC, the negotiator of the agreement, became the first designated carrier (later replaced by Air China) and launched its Beijing–Shanghai–San Francisco–Tokyo–New York service in January 1981. Contrary to the case in China where the allocation of the traffic rights involved no more than some paper work, for the USA, double designation could only partly satisfy the needs of the US carriers that were all very keen to serve the USA–China market. According to Beane (2007), in addition to Pan Am, other major US carriers including Trans World Airlines (TWA), Northwest Orient Airlines (Northwest), United Airlines, and World Airways started to prepare the China services in the late 1970s, anticipating that a diplomatic relationship would be resumed following the historic visit of President Nixon. Pan Am, TWA, and Northwest were even authorised traffic rights by the CAB in 1979. The successful conclusion of the 1980 ASA eventually enabled Pan Am to resume its service to Shanghai and Beijing in January 1981, more than 40 years after its last such operation in 1949.

The politically significant air links did not generate much traffic in the early days of the operations. In 1983, passenger traffic in both directions was only 130,000 and cargo 3,000 tonnes. Due to the low demand, Northwest did not launch its Seattle–Shanghai service on Route B until 1984, while the CAAC did not start its Beijing–Shanghai–Los Angeles (Route B) operations until 1987, being replaced by China Eastern in 1992 with an amended routing from Shanghai–Los Angeles–Seattle and Beijing–Shanghai–Seattle–Chicago as a result of the carrier's successful negotiations with the central government to deploy its newly introduced MD-11 aircraft on long-haul services. By the end of the 1980s, capacity had been only moderately increased to six frequencies per week with no increase in designated carriers.

Discussions in 1992

Sino-US political and economic relations before 1992

From the 1980s till 1992, the USA had two presidents, Ronald Reagan and George H.W. Bush. For Reagan, a pro-Taiwan policy was pursued, which threatened Sino-US relations. The Shanghai Communique of 1982 reassured China of the US' position in the region, which helped restore the relationship, leading to a recovery of bilateral trade. By the time President Reagan visited China in April 1984, the first in-office US president to visit China after the two nations established diplomatic relations seeking to improve the friendship, China was in the fourth year of a row over a large trade surplus with the USA. In the first half year of 1983, US exports to China were $1.1 billion, a 12.4 percent increase over the same period in 1982, while US imports from China increased by 40 percent to $1.5 billion (Chen and Lee, 1984).

Domestically, the Reagan Administration advocated a free competition policy. He allowed the larger incumbent carriers to benefit from a by-standing approach, which enjoyed a state of virtual non-regulation to rebuild much of their lost hegemony (Havel, 2009). The relaxed environment nurtured the overhaul of the structure of the US airlines industry, resulting in bigger airlines such as American, United, and Delta, which had adopted hub-and-spoke systems developing their networks to achieve economies of scale, while start-up carriers delighted the travelling public with new routes and low fares (Havel, 2009). At this time, the US government was unsure about the policy of promoting the liberalisation of international air transport to its trade partners, who had been accustomed to the old economic compromises of bilateralism (Havel, 2009), hence the focus remained on the domestic market.

When George H.W. Bush was elected into office in the later 1980s, his goal was to strengthen the relationship with China so as to usher in a new level of stability (Garrison, 2005). The 1989 event in China challenged Bush's policy, leaving him to press hard to maintain the relationship. In such a complex domestic context, Bush managed to continue the engagement with China hoping to forge a broad role in the architecture of the emerging order in Asia while at the same time countering the Soviet Union. He supported China's economic reform, which, he argued, would eventually lead to political freedom with foreign trade and investment being essential tools to keep China open to the outside world, thus availing an opportunity for the USA to have an influence to work for cooperation, which would be a more reasoned, careful action taken given its long-term interests in China (Garrison, 2005).

By maintaining a balanced relationship with China and forging a continuous senior level dialogue and shifting political controversy to China's economic reform and its achievements (Garrison, 2005), Bush supported China's efforts in joining the WTO and managed to retain China's Most Favoured Nation (MFN) status and successfully improved US market access to China to help Beijing make the reform required for WTO membership.

In terms of air transport, the focus of the US government was on improving aviation relations with EU member states, such as the entrepreneurially inclined Netherlands that signed the first-ever "open skies" agreement with the USA (Havel, 2009). The DoT initiated the idea of an "Underserved Cities Program" in 1989 so as to allow foreign airlines to provide extra-bilateral international services to and from US city-points that did not otherwise receive the identical international non-stop service from US airlines (Havel, 2009), thus encouraging airlines of the EU member states to expand their operations on the cross-Atlantic markets. Consequently, the US carriers' attention was also on the US–EU markets rather than on the Asia Pacific, including China, where public demand was too low to stimulate the major US carriers to explore the market.

Political and economic reform before 1992 in China

Taking a controlled and progressive approach, China's economic reform initially concentrated on enterprise restructuring by consolidating its property rights and by adopting a new enterprise governance structure which stressed enterprise autonomy and incentives (Li, 1994). The enterprises were encouraged to take accountability of their performance and be responsible for their own losses and profits. Due to the strategic importance of Chinese airlines which was regarded as part of the military defence forces, the reform of the industry was almost ten years later than that in other industries, with the first airline being formulated based on individual branch of the CAAC's regional administration in 1987. The CAAC remained to be a regulator responsible for formulating the air transport policies for both domestic and international services as well as a flag carrier operating all routes. Primary efforts were given to steering domestic economic reform of the airline industry, with international operations being considered a subordinate activity, with CAAC consulting and reporting to the MFA to ensure that the negotiation results were to support China's political objectives and that the air links between a third country with Taiwan was of a commercial nature with no political implications.

Agreement reached in 1992

In this context, the China–USA consultation held in 1992 brought about the first relaxation since the conclusion of the ASA. It was agreed that a third carrier would be allowed to enter into the market, which could be either an all-cargo operator or a passenger airline with cargo business to introduce all-freight services to the market. The new designated carrier would be given the right of flying between any points in the USA and any airports that were open for international operations in China via any intermediate points and further to any beyond points with full traffic rights. This was a very innovative and exceptionally relaxed arrangement in the early 1990s, taking into account that there were no all-cargo operators existing in China at all. For the USA, the agreement represented a remarkable achievement in the Asia Pacific, given the fact that it had

TABLE 4.4 Key points in the agreement reached in 1992

Item	Contents
Route A	Beijing–Guangzhou–Shanghai–Tokyo/or another point in Japan–Honolulu–Los Angeles–San Francisco–Chicago–New York v.v
Route B	Beijing–Guangzhou–Shanghai–Tokyo or another point in Japan–Honolulu or Seattle–Los Angeles–San Francisco–Chicago v.v.
Cargo route	Any point in its own country–any intermediate points–any points open for international operations in the other country–any beyond points, with full traffic rights
Number of designated carriers	3 for each side
Weekly frequencies for each side	18 for 1991–1992 20 for 1993–1994 27 for 1995–1996

Source: Adapted from the 1992 ASA amendments.

just concluded an open skies agreement with the Netherlands which allowed free operations between the two countries. In terms of capacity, both sides agreed to increase the weekly frequency to 27 for each side by 1996 through phased-in arrangements compared with the previous six weekly frequencies. In addition, both sides agreed to approve any commercial arrangements reached between the carriers including wet leasing, block seats, and codeshares (Table 4.4).

The 1992 arrangements allowed Evergreen International (later replaced by FedEx) from the US side to start an all-cargo service on the route New York–Chicago–Anchorage–Beijing–Shanghai in September 1993 and Air China to introduce an all-cargo service from Beijing to New York via Shanghai and Anchorage from the winter season of 1993. Traffic grew rapidly for the following two years with both sides being convinced that there was a need for further discussions before the capacity limitation was reached.

Consultations in 1995 and 1999

The years leading to the 1995 and 1999 discussions were very different for both China and the USA. This period was marked by a more dramatic unease in Sino-US political relations due to the economic sanctions imposed on China after the June 4th incident, the US arms sale to Taiwan, the permission of Li Denghui's visit to the USA in 1995, and the bombing of the Chinese Embassy in 1999. The US government had successfully concluded tens of open skies agreements with the EU member states, which resulted in the EU accelerating the pace of unification of its air transport market. As liberalisation had shown its impact on airlines operations as well as the benefits perceived by the consumers, the US government was increasingly determined to promote open skies around the world and, thus, shifted its focus to the Asia-Pacific countries.

The USA's China policy

In his first term, President Clinton tried to define his relationship with China according to their ideological differences, although he shortly abandoned this to adopt a more pragmatic approach instead. Focusing on commercial relations in his China policy rather than the traditional security issues, which was a marked shift from his predecessors, Clinton was convinced that engaging China to promote its economic liberalisation would eventually enlarge democracy (Garrison, 2005). Although this caused some incompatibility between his foreign policy and domestic policy priorities, Clinton managed to balance trade and human rights issues by choosing economy as the central theme. He delinked the MFN with human rights issues in 1994, arguing that the USA had strategic, economic, and regional concerns beyond human rights and that there was nothing contradictory between American ideals and commercial interests in China with his supporters voicing that the commercial relationship would serve as the important first steps to make military dialogues possible. McDonnell Douglas was able to seal a deal with China to provide the country with commercial aircraft worth $1.6 billion, which were later used by China Eastern to launch its first-ever international long-haul route to serve the USA.

Clinton's second term saw a fluctuation in Sino-US relations, a critical time when China was in the final throes of applying for membership of the WTO. A fear appeared that China could emerge as an Asian power, which could be a threat to the US' interests in the region, causing the US ruling military elites to demand a far more unilateral US military and diplomatic policy against China (Head, 1999). They argued that a containment policy would be a more realistic way to deal with the prospect of a powerful China and efforts should be concentrated on slowing down its economic growth and preventing China from upgrading its military capabilities (Khalilzad, 1999). A series of events, including the belated intelligence claims that a Chinese spy obtained US nuclear secrets, criticism about China's threat over Taiwan, the jailing of political dissidents, and a widening trade gap, fueled the tension eroding bilateral relations (Head, 1999). Then came the deliberate bombing by NATO of the Chinese Embassy in Yugoslavia during the war in Kosovo (Sweeney et al, 1999), which infuriated the whole of China, causing major anti-America nationalism across the nation. The Chinese representative accused NATO of carrying out a war crime at an emergency session of the UN Security Council, the Chinese press carried front page stories with photos of the victims of the Embassy bombing, and major cities saw their biggest and most fervent demonstrations in response to the bombing. In Beijing, about 100,000 people invaded the Embassy district and the residence of the US

Consul General in the south-western city of Chengdu was stormed and partially burned (BBC, 1999).

The USA and China broke off diplomatic contact for about four months until discussion on China's entry into the WTO softened relations. Clinton eventually was able to push forward his agenda including supporting China's entry into the WTO, arguing that if China was willing to play by the global rules of trade, it would be an inexplicable mistake for the USA to say no. He stated that China's entry into the WTO was in the national interest and would represent the most significant opportunity that the USA had ever had to create a positive change in China. Through his efforts, Sino-US economic relations expanded to a great extent with China recording in 2001 $54.3 billion in exports to the USA representing a 4.2 percent increase year on year and $26.2 billion in imports, (an increase of 17.2%). (Wang, 2002). The USA had become the second largest trading partner and the largest importer of Chinese products, whereas China was the fourth largest trading partner of the USA (Wang, 2002).

China's view towards the USA

China's multipolarity policy was pursued since the early 1990s, when the Soviet Union collapsed, which dramatically changed the world order. China was convinced that the disappearance of the Soviet Union meant that a multipolar world would be the trend with China being counted as a pole together with the USA and Russia. Deng elaborated in 1990 that

> the situation in which the United States and the Soviet Union dominated all international affairs is changing. Nevertheless, in the future when the world becomes three-polar, four-polar, or five-polar, the Soviet bloc, no matter how weakened it may be and even if some of its republics withdraw from it, will still be one pole. In the so-called multi-polar world, China too will be a pole. We should not belittle our own independence: one way or another, China will be counted as a pole.
>
> *(Deng, 1990)*

When Jiang Zemin took office in 1989, he followed Deng's view towards the USA, believing optimistically that the USA's power would soon decline with the rising of other countries such as Japan, Germany, and China (He, 2009). To implement the policy, significant efforts were given to boost links with other countries and regions in the world although the Sino-US relations remained to be one of the priorities. For example, China established strategic partnerships with Japan, the EU, and Russia, hoping that such an arrangement with other great powers would increase China's balancing capability to countervail the USA (He, 2009). At the same time, to promote its multilateral diplomacy, China proactively participated in international organisations supportting the development of multilateralism. For example, China joined the Asia Pacific Economic Cooperation

(APEC), ASEAN, initiated ASEAN plus China, ASEAN plus Three, and led the Shanghai Five, which became the Shanghai Cooperation Organisation (SCO) (He, 2009).

The fluctuating relationship affected the political exchange between the two countries. There had been no presidential meetings between the two nations for eight years[2] until November 1993, when China's president Jiang Zemin met his US counterpart at an unofficial occasion in APEC in Seattle. The formal meeting between Jiang and Clinton did not happen until two years later in October 1995, when both were attending the 50th anniversary of the United Nations. Jiang's state visit to the USA eventually took place in October 1997, when a joint communique was announced confirming that both parties would adhere to the principles in the Sino-US three Communiques and agree on the objectives, principles, and guidelines governing Sino-US relations in the 21st century (Zhang, 2008).

Nevertheless, the instability in the political relationship in the 1990s did not hinder the dramatic surge in bilateral trade, with China beginning to run a rising trade surplus with the USA. In 1990, China's trade surplus was only $10.4 billion. The figure jumped to $57 billion in 1998, which was only a few billion less than that of Japan, the USA's largest trade-deficit partner at that time (Feng, 2006).

For this period of time, the Chinese government still concentrated its efforts on domestic air transport development, which grew at double digit rates. Although foreign direct investment in China as well as China's exports had increased dramatically, international air transport was not regarded as a strategic sector that could further facilitate and promote the country's export-oriented economy. Instead, it was still considered as a political instrument to support the nation's diplomatic strategy to promote China's presence around the world. The CAAC simply followed the instructions of SC and MFA in dealing with international air transport issues and did its utmost to protect its carriers by being reluctant to lift the restrictions on the air transport markets.

US's push for open skies in Asia Pacific

For international air transport, Clinton himself was a strong advocator of the US "open skies" policy and appreciated the benefits of a more liberal multilateral trade regime. He empanelled the Airline Commission to uncover the structural causes of the parlous financial condition of the US airlines industry, which, in its report, called for the ceiling on inward investment in US airlines to be raised and recommended a multilateral "open skies" replacement for the patchwork of bilateral agreements (Havel, 2009).

The DoT "International Air Transportation Policy Statement" in 1995 was an official declaration of the US government's intention to eliminate the central pillars of the prevailing Chicago system of protective bilateral agreements (Havel, 2009). Open skies had since been aggressively pursued around the globe, with Asia Pacific becoming the primary focus from 1996 after the USA had successfully conquered some European countries since 1992.

Not surprisingly, a pattern of discord had been shown by Asian countries. For example, Japan expressed continuing discontent with the fifth freedom rights which enabled US carriers to combine at Tokyo, traffic flows from a variety of US cities for onward transit to multiple destinations in the Asia/South Pacific market (Havel, 2009). Australia maintained a striking conceptual separation between its highly deregulated and denationalised domestic aviation marketplace and the classic bilateral aerodiplomacy through which it aggressively stewarded its international markets to safeguard the interests of its privatised flag carrier, Qantas (Havel, 2009). Notwithstanding the opposition, the USA was successful in convincing Asian countries to conclude "open skies" agreements using the same tactics of "beggar-thy-neighbour." Following the agreement with Macau SAR in 1996, Singapore, Taiwan, South Korea, and Thailand all agreed to the " open skies" arrangements with the USA through protocol or exchange of notes (Havel, 2009).

Airlines performance

For China, its continuous economic boom in the 1990s had led to a phenomenal surge in the demand for air transport. In 1999, a total of 60 million passengers were carried compared to 11.5 million in 1987 (Meyer, 2002), representing a 5.3-fold increase year on year. However, the explosive growth had led to a spate of air accidents leaving 642 casualties between 1989 and 1994 (Meyer, 2002), which resulted in tight controls in terms of market entry and pricing being imposed by the Chinese government. The financial crisis in Southeast Asia in 1997 also had a severe impact on the profitability of Chinese carriers, which collectively lost $294 million in 1998 and another $200 million during the first six months of 1999 (Meyer, 2002). The three carriers in the Sino-US market, for example, Air China, China Eastern, and China Southern[3] were only able to carry 327,540 passengers and 62,700 tonnes of air cargo collectively in 1998 with an average 56.9 percent load factor on all their international routes (Meyer, 2002).

In contrast, the US airlines had experienced a golden period of growth and prosperity after enjoying two decades of deregulated operational environment in the domestic market. The powerful sales and marketing tools such as CRS and FFP as well as strategic hub-and-spoke network system had underpinned the strength of US carriers in both domestic and international markets. The open skies agreement concluded with 31 partners by 1998, five of which were Asian economies (US Department of State, 2010), had created a liberal environment for US carriers to expand aggressively into global markets, especially in Europe. Codeshare arrangements had facilitated the US carriers bypassing the stringent bilateral restrictions to further optimise their hub-and-spoke networks and exploit market opportunities.

In addition, the US carriers had enjoyed handsome profits for several years. For example, United and Northwest, which had been operating to China for

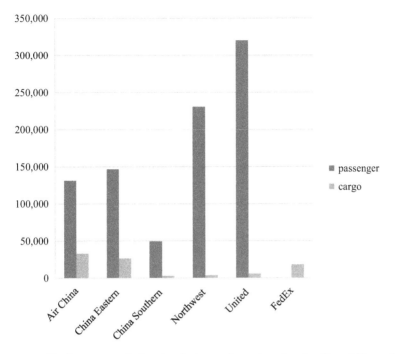

FIGURE 4.8 Performance of the six designated carriers in the Sino-US market in 1998.

Source: www.caac.gov.cn.

over a decade, had remained highly profitable into the late 1990s, posting annual profits in 1999 of $1.24 billion and $300 million respectively (Meyer, 2002). The two airlines collectively carried 550,800 passengers, 68 percent more than their Chinese counterparts. The consolidator FedEx reported $442 million profits in 1999 with an average cargo load factor of 64.2 percent on all its international flights in 1998 (Meyer, 2002). In the Sino-US market, the three US airlines[4] shipped a total of 27,400 tonnes of cargo, though only 44 percent of the total market. Figure 4.8 shows the performance of the six designated carriers in Sino-US market in 1998.

The agreement reached in 1995

The meeting in 1995 led to a protocol signed between the two parties, which allowed the designated carriers to enter into code-share arrangements to facilitate passenger travel. The key points are summarised as follows:

• A third combination carrier for China would be allowed due to the absence of an all-cargo operator as well as one more gateway city in the USA with full traffic rights for all Chinese carriers;

- An increase of weekly frequency from two to four was agreed for the US' cargo operator; and
- Codeshare arrangements between the designated carriers were encouraged.

The amendments enabled China Southern to become the third Chinese combination carrier which selected Los Angeles as its first US destination and FedEx to increase its operations to four weekly frequencies. The amendments also allowed direct services without any stopovers, such as at Tokyo, between the two countries, resulting in Northwest launching a direct Detroit–Beijing service later in 1996. The carrier deliberately chose NW88 as its flight number to attract Chinese passengers, who culturally believe that the number eight would bring great fortune and good luck to one's life. The reduced flight hours (from 15 hours via Tokyo to 9 hours) by flying across the Arctic above Russia boosted the enthusiasm of Chinese passengers to visit the USA, who were convinced that the USA was quite reachable.

Benefiting from the phenomenal growth in bilateral trade and passenger traffic, carriers of both sides reported a shortage of capacity by the end of 1998 when both sides had used up their entitlement of 27 weekly frequencies. To facilitate the airlines to meet the market demand, the two parties sat together in April 1999, which resulted in a protocol that considerably relaxed the restrictions on operations.

Expectations of both sides before 1999 and the arrangements reached

Encouraged by its achievements in open skies arrangements with five Asian economies as well as those in other continents, the US government had a high expectation out of the consultation with China (Meyer, 2002), with the hope of removing barriers for the ambitious airlines such as Delta, American, Polar Air Cargo, and UPS to enter into Asian markets, with China being one of their key targets. In contrast, the Chinese side was not enthusiastic about opening up its air travel market further for free competition, given its loss-making airlines and embarrassing safety record over the past decade (Meyer, 2002).

The divergent expectations did not prevent both parties from reaching an agreement despite the negotiations being tough. The protocol concluded was considered the greatest step forward towards liberalisation of Sino-US aviation trade since 1980 (Meyer, 2002). It authorised a fourth carrier from each side to enter the market beginning in April 2001 and another 27 frequencies (totalling 54 for each side representing a 13.5 fold increase compared with the four weekly frequencies of 1980 or twice as many as that agreed in 1995) through phased-in arrangements. It also granted the airlines the freedom to originate their combination services (on Routes A and B) from any city of their choice in their respective home countries and land at a few more destination cities in the other party's home country, bringing the total number of gateway cities in the USA served by the Chinese carriers to 12 and that in China to 5. Table 4.5 summarises the amendments reached in 1999 compared with the agreement of 1980.

TABLE 4.5 Summary of the 1999 agreement and the comparison with 1980

Item	Contents	Compared with 1980
Routes A and B for both sides	5 gateway cities in China, 12 in the US	2 in China, 5 in the US
Route C/Cargo route	No limitations in terms of airports to be served by the other party	No dedicated cargo routes available
Number of designated carriers for each side	4	2
Weekly frequencies for each side	54	4
Codeshare arrangements	Yes	Not permitted

Source: Adapted from the 1980 and 1999 ASA amendments.

Implementation of the protocol and reaction of the industry

Again, the amended agreement could not satisfy the needs of the major US carriers who applied for USA–China traffic rights. The incumbents such as United and Northwest wanted additional operations to achieve economies of scale while the prospective entrants wanted the opportunity for new ventures. For example, Delta lobbied for a daily non-stop service between New York and Beijing (Meyer, 2002), while American wanted to launch a brand new service out of its Chicago hub so as to allow it to create links for some 80 US communities to China. It argued that the services could cover both passenger and cargo operations which would help enhance social as well as economic ties between the two societies (Beane, 2007). Polar Air Cargo and UPS both requested all-cargo services to China, claiming that such provisions would help greatly improve the employment of US industries (Beane, 2007). Eventually, the DOT chose UPS as the designated fourth carrier in the USA–China market, largely due to the strong support from more than 350 members of Congress, various state governors, and the intense lobbying of the Teamsters union (Beane, 2007). A six weekly all-cargo service was launched in April 2001, thus ending FedEx's monopoly in the Sino-US cargo market (Beane, 2007). The DOT also split the remaining slots between the incumbents including Northwest, United, and FedEx, enabling them to provide more frequent services between the two countries.

Contrary to the fierce competition for traffic rights among US carriers, the Chinese side went through a quiet process in introducing China Cargo Airlines (CCA), an all-cargo operator affiliated to China Eastern, to the Sino-US market as the fourth carrier. This was understandable since there were no other carriers existing in the market that were licensed by the Chinese authority to operate long-haul international routes. The incumbents such as Air China remained eligible for all-cargo operations under its capacity entitlement, while China Eastern and China Southern were able to carry cargo on their passenger aircraft.

Although codeshare arrangements were agreed and encouraged in 1992s amendments, regrettably, none was concluded between the designated carriers. One of the reasons was that neither the Chinese government nor the Chinese carriers were prepared to discuss any codeshare arrangements as they did not feel comfortable nor confident about this kind of commercial deal since they were uncertain of what codeshare meant and what the outcome would be for them.

In the following years, more engaging activities were organised to address the industry concerns. Analytical commentaries were published in industry journals to explain what codeshare was. US solicitors and aviation consultants were invited to give lectures on how the codeshare would work with the benefits and pitfalls being elaborated. Eventually, almost six years after the initial introduction of the codeshare concept, the Chinese side was prepared to negotiate codeshare deals with their US counterparts, leading to agreements reached between almost all designated carriers in 1998. Northwest agreed a wide range of code-share arrangements with Air China, which gradually developed into an exclusive cooperation between them. China Eastern code shared with American, and China Southern collaborated with Delta.

A milestone protocol reached in 2004

The USA's China policy

When George W. Bush took over the presidency from Clinton, he, in the early days, repeated what Clinton did by breaking with the Republican tradition to take China as a strategic competitor. Soon afterwards, Bush changed his tone from one of "containing them" to "engaging a friend," shifting his China policy to allies after the 9/11 event. Over the following years, Bush's China policy was viewed as one rare bright spot in the 21st century featuring both unprecedented bilateral cooperation on shared concerns, such as terrorism, the global financial crisis, and economic issues such as trade imbalances, as well as competition for regional influence (Shi, 2009). China's accession to the WTO helped involve the country in a web of interdependence with the rest of the world, in particular with the USA, which imported $1.89 billion in 2001, a 27.2 percent increase year on year according to statistics by the Ministry of Commerce of the People's Republic of China. On the other hand, China's continuous export growth to the USA brought the US trade deficit to a new level, with antagonism towards China flaring in the USA (Lynch, 2005). Sino-US strategic economic dialogue (SED) aimed at maintaining a direct communication at the senior level and USA–China Aviation Cooperation Program (ACP) aimed at promoting technical, policy, and commercial cooperation between the aviation sectors of the two countries were launched in 2004, which produced a positive effect on aviation with closer links and exchange activities.

China's US policy

Since the late 1990s after the Kosovo War, China had gradually changed its perception of multipolarity, being convinced that the USA would remain as the sole power of the international system, which would still pose a threat to China's security (He, 2009). China adjusted its US policy by choosing to be accommodating and not confrontational, stating that the military gap between the USA and all other powers remained constant and was expanding such that the US-led unipolar world would remain dominant instead of declining (He, 2009). Being quick and flexible in dealing with conflicts with the USA, China manages to maintain a good relationship, with economic development and trade promotion being considered as the best strategy to deal with the USA. In October 2002 during Jiang's working visit to the USA, both parties recognised that China and the USA should improve the bilateral dialogue and coordination on important international issues and to expand the cooperation in all areas so as to strengthen further the constructive working relationship.

Hu and Wen's Administration after 2003 has followed the same strategy when dealing with the USA, although at the same time China has reinforced its relations with its peripheral states, those in Africa, South America, and the Middle East. China believes that peripheral states should take priority in its diplomacy despite big countries still being the key targets. The USA is critical for China's peaceful rising and a strategic and constructive relationship should be maintained based on mutual benefits.

To implement the strategy, in December 2003, soon after Hu and Wen took office, Premier Wen proposed five principles with respect to the development of Sino-US fair trade and economic cooperation, which included:

a   Creating a win-win situation for mutual benefit;
b   Taking development as a priority to resolve any disputes through economic and trade cooperation;
c   Optimising the collaboration mechanism for timely and effective communication and discussion;
d   Coordinating and discussing fairly to seek for common ground; and
e   Not politicising economic and trade issues.

To ensure that China has a bigger role to play in managing the Sino-US relations, Chinese President Hu Jintao, in April 2006, when launching the SED, further elaborated a six points proposition, three of which were:

a   Improving mutual understanding to expand the common ground so as to develop a long-term, stable, constructive, and cooperative relationship;
b   Seeking for opportunities to strengthen the bilateral economic and trade cooperation;
c   Respecting each other, appreciating, and properly addressing the differences.

With the above doctrines as guidelines, China has pursued its US policy in a wider context, especially in the realm of trade and economic development, including currency stability, energy sufficiency, and the projection of Chinese "soft power."

In terms of air transport, the US "open skies" policy had become widely accepted with around 90 nations accepting such agreements with the US during Bush's tenure. The US also concluded with the EU the first stage open skies deal in 2007, which served as a model for other country-pair negotiations. For the USA–China market, the 2004 Protocol was widely regarded as a landmark in removing the economic restrictions constraining the commercial operations of air transport between the two nations.

Market development between 1999 and 2004

The years following the 1999 Protocol retained the momentum of continuous growth of trade and traffic between the two countries with 2001 marking a historical milestone when China gained WTO access. China attracted over $100 billion investment from the USA in 2003 involving more than 40,000 projects and more than 180,000 Chinese students studying in the USA. Travel in both directions increased by an average of 12 percent over the four years, with American arrivals in China increasing by 15 percent while Chinese visitors to the USA rising by 8 percent. However, the 9/11 terrorist attack had a severe adverse impact on air travel which resulted in the tight control of US visas and stringent security measures, causing a drop of 5.7 percent in 2002 of Chinese tourists heading for the USA. Before the market was able to recover, SARS in 2003 caused another loss of passenger traffic which saw 27 percent decrease for the first seven months of 2003.

Nevertheless, the 1999 amendments created a framework for airlines to meet the market demand. The US carriers were quick in taking actions to capitalise in the market. Northwest Airlines launched its twice-weekly Detroit–Shanghai service in the spring season of 2000 and achieved an average 85 percent load factor for the first few months, contributing to its extraordinary performance in the Asia Pacific region, where nearly 23 billion RPM was generated during 2000 (Meyer, 2002). United reported a similar outstanding performance in the region with an average 70.8 percent load factor producing 11.5 billion RPM for the first six months of 2000. As the market continued booming, the four US carriers, namely United, Northwest, FedEx, and UPS, soon found themselves restrained by the capacity restrictions for further expansion in the market, as they had collectively used up all the 54 weekly frequency entitlements in 2001, which was only more than one year after the 1999 amendments. Finding themselves in a constrained circumstance, carriers such as Northwest went to their Chinese counterparts for help in attempts to conclude any kind of commercial arrangements so as to allow the carrier to virtually expand their operations into China, although their efforts amount to much.

The 9/11 tragedy resulted in US airlines making a significant loss, leaving quite a few major carriers such as United and Delta seeking Chapter 11 bankruptcy protection. Faced with the sluggish domestic market, the US carriers recognised that great opportunities existed in international markets which would help compensate for the demand loss. A well-noted trend had been that the major US carriers have deliberately focused their route planning efforts on more profitable international routes, such as Asia Pacific, Commonwealth of Independent States and Middle East, which would produce higher yields (Compart, 2008). For example, United reduced its capacity in the North American market by nearly 14 percent between Aug 2004 and Aug 2005 but at the same time increased its operations to China by adding more frequencies and destinations, increasing its market share from around 25 percent in 2000 to 49 percent in 2005 (CAAC, 2005).

In contrast to the speedy expansion of the US carriers, the four Chinese carriers, namely Air China, China Eastern, China Southern, and CCA, were quite cautious in increasing capacity and destinations compared with their US counterparts. Air China managed to launch a thrice weekly non-stop service between New York and Beijing in September 2002 and a twice weekly cargo service between Beijing, Oregon, and Portland in October 2002 while China Southern started thrice weekly cargo flights between Guangzhou and Los Angeles at the same time (Meyer, 2002). One of the reasons for the reluctant reaction was that Chinese carriers found it difficult to make money from the international operations compared with their lucrative domestic market where a high demand has remained for decades. Data available showed that Chinese carriers were only able to produce approximately 10.3 billion RPK collectively in 2000, less than half of that generated by Northwest's operations in the Asia Pacific region for the same period (Meyer, 2002). In 2001, the three flagship carriers made a total loss of less than $10 million, engendering cautious responses when seeking to increase capacity. The slow implementation did not bring the total frequencies of Chinese carriers to its full entitlements. By the summer/autumn season of 2003, the Chinese passenger carriers had only used up 44 weekly frequencies, 81 percent of their total entitlements. Figure 4.9 is the split of passenger traffic between the two countries from 1999 to 2002, which demonstrates that US carriers were carrying more traffic than their Chinese counterparts.

In addition to the scheduled services, carriers of both sides also requested additional services to meet the special demands in certain periods of the year. According to the statistics of the Chinese authority, in 2003, 114 additional frequencies were flown by US carriers to Chinese destinations representing 2.2 more frequencies per week and a total of 94 extra flights were operated by Chinese airlines to US destinations, indicating 1.8 more weekly frequencies.

Contrasting to the passenger market, where the US carriers were keen and able to capitalise while the Chinese were reluctant and passive in competing, competition in the air cargo market was more equitable. Airlines of both sides

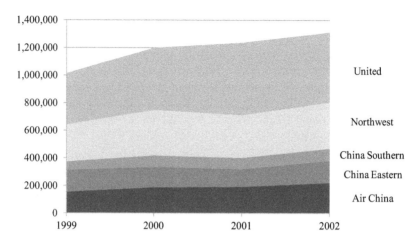

FIGURE 4.9 Passenger traffic between the two countries from 1999 to 2002.
Source: www.caac.gov.cn.

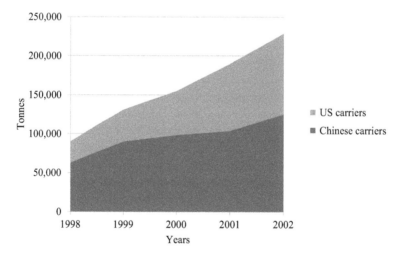

FIGURE 4.10 Air cargo volumes between 1998 and 2002.
Source: www.caac.gov.cn

were not only eager to increase frequencies but also proactively competing with each other for a bigger market share. This was reflected in the operational volume as well as the load factors. As a result of the continuous booming of bilateral trade, the total volume of air cargo between the two countries increased to 229,000 tonnes in 2002 from 90,000 tonnes in 1998, representing a 26.7 percent increase year on year (Figure 4.10). When examining the performance of each carrier, the statistics available shows that UPS and China Eastern were able to gain an increasingly bigger market share while Air China's dominant position was dismantled (Figure 4.11).

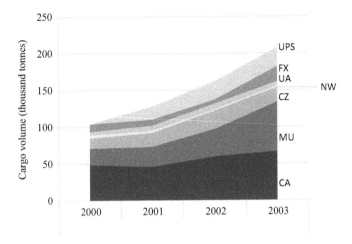

FIGURE 4.11   Split of air freight carried by the designated carriers between 2001 and 2003.

Source: www.caac.gov.cn.

Carriers of third countries competing in the Sino-US market

Since the beginning of the 21st century, the Sino-US market has become more lucrative. While the US carriers were restrained from expanding capacity to capture the market demand as a consequence of the regulatory controls in place and with the Chinese carriers failing to utilise their full entitlements leaving a handful of weekly frequencies unused, airlines from third countries capitalized on the market opportunities. Taking advantage of their liberalised arrangements with both China and the USA, these third-country carriers were able to respond to the market demand. During this early period, it was customary for the majority of freight between China and the USA to be shipped not by the designated carriers of both sides but rather those of third countries airlines utilising the fifth or sixth traffic rights between China, the USA, and their home countries, as well as non-designated US carriers such as Evergreen International and Polar Air Cargo. For example, Singapore Airlines operated four weekly services to Anchorage, Chicago, and Los Angeles via Nanjing and Xiamen in China with the fifth traffic rights. Korean Air and Japan Airlines Cargo operated between China and the USA via Seoul and Tokyo by using the sixth traffic rights.

The significant leakage of cargo not only caused the loss of revenues for designated carriers but also adversely helped those third-country airports to develop into hubs to compete against Beijing, Shanghai, and Guangzhou, which are China's biggest three airports that have been also ambitious to develop into hubs in Asia. Table 4.6 summarises the carriers and their total cargo capacity operating between China and the USA in the summer season of 2003, which indicated that

TABLE 4.6 Cargo carriers competing in the Sino–US market (summer season in 2003)

Airports	Carriers operating at the airport	Weekly frequency in total	Destinations in the US
Hong Kong	Northwest	68	Anchorage
	Polar Air Cargo		New York
	FedEx		Chicago
	Evergreen International		Los Angeles
	Cathay Pacific		
	Japan Airlines Cargo		
Seoul	Korean Air Cargo	97	Los Angeles
	Asiana Airlines		New York
	Northwest		Chicago
	Polar Air Cargo		San Francisco
	FedEx		
	UPS		
Tokyo	Northwest	50	Anchorage
	Japan Airlines Cargo		New York
	Japan Airlines		Los Angeles
	Lufthansa		Chicago
	Polar Air Cargo		
Beijing	Air China (Cargo)	19 (respectively)	New York
Shanghai	China Cargo		Los Angeles
Shenzhen	China Southern		Portland
			Chicago
Beijing	FedEx	20 (respectively)	Memphis Anchorage
Shanghai	UPS		New York
	Northwest		Dallas
Nanjing	Singapore Airlines	4 (respectively)	Anchorage
Xiamen			Chicago
			Los Angeles

Source: Compiled by the authors from various sources.

the arrangements between China and the USA reached in 1999 were no longer able to satisfy the market demand, which necessitated the urgent need to increase the direct services and capacity.

Preparation by both sides

The growing market demand and inadequacies in the capacity encouraged both governments to take further actions to expand the arrangements. The two agreed to resume discussions in late 2003. By the end of 2003, the USA had concluded 59 open skies with its partners, almost twice as many as it had in 1998 (31 open skies by the end of 1998). Among them, ten were Asian economies including Taiwan, Singapore, Malaysia, Brunei, New Zealand, Pakistan, Uzbekistan, and South Korea. These open skies deals facilitated the US carriers developing their networks in the Asia Pacific region, thus strengthening their global market leading positions. The USA was convinced that it was the right time for China to lift the market entry barriers to enable free competition since it had been three years with the WTO.

In terms of the Sino-US market, as China's overall trade volume was projected to increase substantially in the coming years, the aviation sector would be expected to play a pivotal role in the evolving Sino-American trade relationship. A DoT study had shown that air freight had been the fastest growing segment of the US cargo industry, hence, the expanding aviation opportunities between the two nations would mean that more US airlines, businesses, and travellers could take advantage of the growing trade between the two rapidly expanding economies (Hong Kong Trade Development Council, 2004). The Chinese market would represent the real long-term opportunity for the US airline industry and open skies with China would allow the US carriers to further capitalise in the market. With the above objectives in mind, the US side wanted to achieve the following:

- To increase the number of designated carriers for passenger operations;
- To separate the passenger capacity from the cargo so as to allow unrestricted cargo operations;
- To allow codeshare arrangements in a third country;
- To allow the US carriers to set up operational hubs in China and change aircraft type enroute of a flight; and
- To allow the US carriers to conduct self ground handling and set up offline offices.

Acknowledging the "wants" of the US side, the Chinese government had its own considerations. China wanted to open up the Central, Western, and Northeastern regions of the country to support the central Government's call for "Western Development" and "Northeast Regeneration" campaigns. The CAAC was in a position to consider the opening up of cargo operations to support the national strategy of an "export-oriented economy". At the same time, the CAAC recognised its obligation to remove market entry barriers to comply with the WTO commitment. However, as to how to remove these barriers, CAAC seemed to have no specific strategies and tactics, nor any contingency measures in place to safeguard the operations of its carriers.

While undertaking the consultation with airlines and airports, the CAAC received divergent feedback. Shanghai Airlines and Hainan Airlines wanted to become new entrants to the market, pointing out that the current Chinese incumbents had failed to utilise all the capacity entitlements, which was truly a waste of resources. Guangzhou airport would welcome as many new direct flights to the USA as possible to support its hub strategy to compete with Hong Kong and to facilitate the local economic growth. Hainan Provincial government requested the island be granted full third, fourth, and fifth freedom traffic rights on the Sino-US markets so as to facilitate the development of the tourism industry of the island.

Contrary to the positive views expressed by the prospective entrants, the incumbents such as Air China, China Eastern, China Southern, and China Cargo strongly opposed any increase of the number of designated carriers and capacity, and did not agree to the codeshare arrangements with a third-country carrier. They argued that their international long-haul operations, especially to the USA, were loss-making services inclined to be affected by any external occurrences, such as the 9/11 terrorist attacks, SARS, and the US visa and security controls. They were disadvantaged in winning passengers, in particular, the high-yield business travellers, due to the stringent visa regime imposed by the US government for foreigners. They demanded a practical mechanism to be in place before any increase of the number of designated carriers and capacity to be agreed and to address the US visa issue which constrained demand of Chinese passengers travelling to the USA. Codeshare arrangements in a third country should not be agreed either. They pointed out that such an arrangement would only allow the US carriers to develop new hubs in the region, such as South Korea, to strengthen their operations in the East Asian region, thus only benefiting the US carriers who had used up their capacity entitlements in the market.

Protocol agreed

With different objectives on the agenda, several rounds of discussions were conducted. In June 2004, the two delegations met for the fourth time in Washington with the aim of striking a deal. The US delegation was headed by Marianne Myles, its Director of Office of Aviation Negotiation in the DoS and was composed of 38 members including officials from the DoS and DoT, as well as representatives from airlines, airports, and industry associations. The Chinese delegation was led by Wang Ronghua, Director General of the Department of International Affairs & Cooperation of CAAC and comprised 26 members including officials from the CAAC and representatives from airlines.

After several days of tough negotiations, a protocol was eventually agreed with the following key points being agreed:

- To increase the number of designated carriers from 4 to 9 by 2010;
- To increase the capacity from 54 to a total of 249 flights[5] weekly for each side by 2010 through a phased-in arrangement (a total of additional 195 weekly flights for each side, with 111 by all-cargo carriers and 84 by passenger airlines). 14 weekly frequencies would become available for new US passenger services as of 1 August 2004. Of the 249 frequencies, up to 133 could be introduced before 2007.
- To open up cargo operations and to allow US carriers to set up cargo hubs in Chinese territory;
- To allow both domestic and third-country codeshare arrangements between the designated carriers of both parties;
- To relax restrictions on commercial activities in a staged manner, including the setting up of offline offices, change of gauge and self-ground handling, etc.

(Source: Adapted from the 2004 Protocol)

The Protocol was recorded as being novel in many respects:

a   It was the first comprehensive amendment of its kind to date which radically liberalised the original ASA signed in 1980 in terms of market entry, cargo operations, codeshare arrangements, ground handling, and offline services;
b   It was the first time in China's aviation history that foreign carriers were allowed to operate to any destinations in the Western, Northern regions, and Hainan Island without any restrictions in frequency, although a few cities such as Urumqi, Xi'an, Chengdu, and Chongqing had been open for limited international operations for quite some time;
c   It was the first time that foreign carriers were allowed to set up a cargo operational hub with full traffic rights in the territory of China. In addition, it permitted a nearly five-fold increase in capacity over the next six years and agreed that pricing for both passenger and cargo operations could be decided by market forces. The Protocol also outlined a timescale for further discussions which would be held in 2006 so as to review the implementation and consider the next steps of opening up the market.

Reaction of stakeholders and implementation of the Protocol

The Protocol was lauded widely by the international aviation community, especially by the US government and industry for effectively creating a regional open skies within China (Beane, 2007). With an estimate of $12 billion additional revenues for US carriers over the next seven years, the Protocol was welcomed as a milestone by the US DoS with Colin Powell making a supportive announcement. It demonstrated that the commitment of the US government to Sino-US economic relations had brought substantial consequences and would bring benefits to US industry, entrepreneurs, and consumers. Norman Mineta, the then Secretary of the DoT, claimed that the agreement recognised the key role played by commercial aviation in Sino-US bilateral trade relations and was a landmark in establishing an international air transport system that would satisfy the needs of the global market.

The liberal arrangements enabled the US DoT to grant most applications from its major passenger carriers to fly or increase capacity to China, although competition for the 111 cargo flights was extremely intense (Caijing Daily, 2004). For instance, United applied for a direct service between Chicago and Shanghai and Northwest intended to launch a Detroit–Guangzhou service with a stopover in Tokyo where its Asian hub is located. Delta, American, and Continental were all satisfied as newly designated carriers though two low-fare carriers Hawaiian Airlines and North American Airlines also applied for traffic rights without success. To make sure that their applications were granted, both American and Delta set up specific websites where people could log on to support the bids for their China routes by sending messages directly to the DoT (Beane, 2007). The campaign proved to be very successful. American was able to cite the support of 26 senators, 78 house representatives, 7 governors, 24 mayors, and 38 airports in favour of their Dallas–Beijing application while Delta and Continental were able to generate more than 10,000 employee letters to back their China services.

Competition for the right of cargo operations was more fierce than for passenger services. The DoT revealed that it had received more than 200 applications for new flights to China (Beane, 2007). UPS planned to add six more cargo flights to China while FedEx wanted to set up a cargo hub in Guangzhou. Polar Air Cargo also endeavored to launch a new all-cargo service to China. The expanded capacity and designation limits enabled all of these requests to be satisfied.

Unlike the overwhelming endorsement gained from the US side, the Chinese reaction was a mixed. The CAAC was pleased to see that it was able to address the imbalance of the market operations in such a way as to satisfy the market demand through proactive discussions with its counterpart and come to an agreement. In so doing, the CAAC was able to embrace opening up of the country to global markets with an innovative approach, hence enhancing the bilateral economic relationship between the two countries to a higher level. With the mounting pressure from local Chinese municipal governments (Caijing Daily, 2004), CAAC was able to support local economic growth by allowing the establishment of the direct air links with the USA, one of China's top export destinations, and to encourage and retain foreign direct investment to central and western China (Caijing Daily, 2004).

Some Chinese carriers welcomed the relaxed arrangements. Shanghai International Cargo, Yangtze Express, and Hainan Airlines were pleased with the Protocol as they were granted Sino-US route licences without much opposition from their peers. They soon launched their cargo services to Dallas and Portland, thus, increasing Chinese carriers' cargo destinations in the US territory in addition to the airports at Anchorage, New York, Chicago, Los Angeles, San Francisco, and Seattle.

The Chinese incumbents responded with anger, panic, and nervousness. They bitterly criticised the CAAC for being unable to protect them effectively by not giving adequate attention to accommodate their needs. They claimed that their views were not taken into account responsibly, nor were their comments on the US proposals considered properly. Their suggestions and counterproposals were just ignored. They not only had no plans to launch any new or additional flights but were extremely concerned about the viability of their existing services. Having such relaxed arrangements with the highly competitive US rivals caused some consternation about their ability to compete effectively given their current fleets, scale of operations, and their high operating costs. Of the 54 weekly flights allowed under the 1999 agreement, Chinese carriers had only run 48 before the 2004 negotiations. Air China operated 22 of the flights. China Southern and China Eastern each ran seven US-bound flights, and China Cargo Airlines operated the remaining 12. With 249 frequencies, the question remained to be asked what benefit the Chinese carriers would exact from the 4.6 fold increase in capacity. "In terms of cargo, China has less than 20 aircraft that can handle trans-Pacific routes," noted one senior industry executive, "how can we compete against America, which has more than 1,000 such planes (Caijing Daily, 2004)?"

The 2004 Protocol had a far-reaching impact on the air transport market for China's peripheral neighbours. South Korea and Singapore reacted by assessing the impact that this arrangement would bring to their carriers, while the Taiwan authority urgently advocated a direct link across the straits so as to dilute the adverse impact on its market as a result of the new arrangements.

The consultation in 2007

The continuous growth in bilateral trade generated a phenomenal increase in both passenger and cargo traffic. The expanded arrangements in 2004 provided opportunities for carriers to capture the market demand. Immediately after the consultation, the six[6] US carriers including the three that were just recently received traffic rights launched a total of 78 new flights (29 passenger and 49 cargo services). By the winter season of 2005/2006, the seven US carriers were operating 114 weekly frequencies with a split of 56 passenger flights and 58 cargo operations. One year later in 2006, the seven US carriers were flying 132 weekly services with 63 passenger and 69 cargo flights, representing 99 percent utilisation of their full weekly entitlements (133 weekly frequencies) to the traffic rights.

Contrasting with the proactive and aggressive action of the US airlines, the Chinese carriers were conservative in implementing the Protocol. Although both Shanghai Airlines and Yangtze River Express were awarded the traffic rights, neither of them launched any services until July 2006, almost two years after the conclusions of the Protocols. In the winter season of 2005/2006, the four Chinese incumbents were operating a total of 63 weekly frequencies with 31 passenger and 32 cargo services. Nevertheless, the Chinese incumbents managed an additional seven frequencies in the Sino-US markets after 2005, with Air China[7] starting three weekly Beijing–Los Angeles services and three weekly Beijing–New York services, and China Southern providing an additional two weekly Guangzhou–Los Angeles flights. In terms of cargo operations, the Chinese incumbents only increased four frequencies after the 2004 Protocol compared with the 49 new cargo services by the US side, which was less than one twelfth. As Air China separated its cargo operations from the passenger business, Air China Cargo became independent and obtained a separate route licence from both the CAAC and the US DoT for the Sino-US market and launched a thrice weekly all-cargo service on its existing routes. The other new service was added by China Cargo which only managed to increase just one frequency per week. By the winter season of 2006/2007, the seven Chinese carriers only operated a total of 76 weekly frequencies with 39 passenger and 37 cargo services, representing 57 percent of utilisation of their entitlements of weekly frequencies (Figures 4.12 and 4.13).

The weekly frequencies operated reflected the performance of the carriers. In 2004, 1.13 million passengers were carried in both directions by the designated carriers with the Chinese side capturing 580,000, taking 51 percent market share. In 2005, the Chinese side carried 680,000 passengers of the total 1.55 million, with their market share dropping to 43.8 percent (Figure 4.14). In terms of cargo traffic, in 2004, of the total 308,000 tonnes of air freight carried by all the designated carriers, 51.9 percent was carried by Chinese airlines. In 2005, of the total 450,000 tonnes of cargo, only 42 percent (190,000 tonnes)

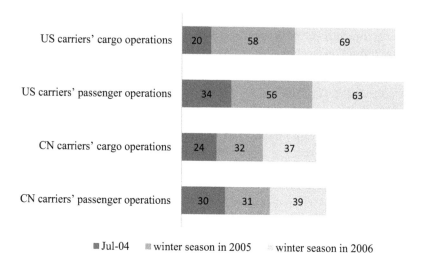

FIGURE 4.12 Weekly frequencies of designated carriers of both sides from July 2004 to the winter season of 2006/2007.

Source: www.caac.gov.cn.

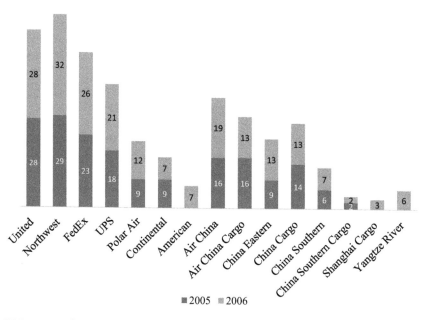

FIGURE 4.13 Comparison of weekly frequencies of the designated carriers operating in 2005 and 2006.

Source: www.caac.gov.cn.

was carried by the Chinese side, with their market share reducing almost 10 percent (Figure 4.14). An examination of the performance of individual carriers revealed that Northwest experienced the biggest increase of market share while the Chinese carriers remained flat (Figure 4.15).

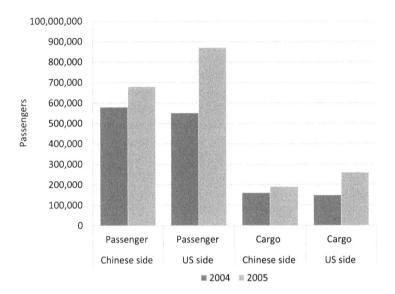

FIGURE 4.14   Traffic carried by carriers between the two countries from 2004 to 2005.
Source: www.caac.gov.cn.

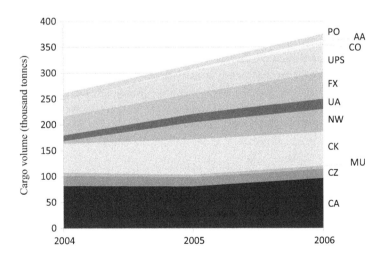

FIGURE 4.15   Cargo volume carried by individual designated carriers between 2004 and 2006 (in thousand tonnes).
Source: www.caac.gov.cn.

Preparations leading up to 2007 consultation by both sides

The sharp contrast of performances as a result of implementing the 2004 Protocol came to the attention of both authorities, who decided to sit together in April 2006, one year earlier than what was committed to in the 2004 Protocol, to review market conditions and discuss the possibilities of further removal of the restrictions on operations between the two parties.

The USA wanted to keep the momentum in 2004, aiming at achieving a full open skies arrangement with China, being convinced that a relaxed operational environment would bring benefits to both airlines and customers, as demonstrated by the outstanding performance from implementing the 2004 Protocol. By this time, the USA was more comfortable in dealing with Asian countries after having successfully concluded 70 open skies deals with its global partners by the end of 2005 with a few more Asian countries being added to the list including Indonesia, India, and Thailand. It believed that it could achieve its ultimate objective by pushing hard with China. Furthermore, the USA held that a more liberalised arrangement of air transport between the two countries would not only help its carriers to expand into the Asia/China market, where they were in a better position to compete, but also support the export of US high technology and the goods of high value to China and help to resolve the imbalance of bilateral trade, of which the USA was in deficit. In addition, it would facilitate the flow of people who would in turn facilitate the promotion of the services industry between the two countries.

The US industry strongly backed its government's proposal. The 2004 Protocol provided great opportunities for them to capture more traffic with a growing market share and strong market presence. They put forward aggressive plans to expand their China operations which included increase of new designations and capacity, and the removal of the restrictions on cargo operations and conditions attached to change of gauge. For example, Delta wanted to be a designated carrier in 2007 and launch services from Atlanta to Beijing and Shanghai while United wanted a Washington–Beijing service. American, Continental, and FedEx all wanted to increase frequencies to Beijing, Shanghai, and Guangzhou. The US airports and local authorities also voiced their interest in having more direct flights to China, such as the Guam Islands which wanted to have as many Chinese visitors as possible to boost their tourism industry and local economic growth.

At this time, the Sino-US relations had been ushered into a new era. The two parties had declared a strategic partnership in 2006 with frequent exchange of visits of senior-level officials. The SED was established in December 2006 with an aim of bringing together the very senior officials to one forum to discuss a wide range of topics including air transport relations. China had replaced Japan to become the USA's second biggest trade partner in 2006, although the trade deficit had grown bigger.

In line with their overall objectives and taking into account the requests from the industry stakeholders, the US side was confident in the groundwork for a new agreement to be reached.

> We want to reach a meaningful agreement about having full and open liberalisation of aviation. I believe we can reach such an agreement by May at the next SED event and hopefully implement the agreement at the end of the year,
>
> (*Payload Asia, 2007*)

said Mary Peters, the US Transport Secretary, during her first visit to Beijing in April 2007. At the least, the USA was prepared to put the "basic framework" of an "open skies" agreement in place at this round of discussions so as to achieve a full range of open skies in March 2010. To ensure that open skies was achievable as planned, the USA wanted to have a transitional arrangement that was able to pave way to the full liberalisation of the bilateral market. To this effect, for the forthcoming negotiations, the USA wanted to remove almost all the restrictions on the operations including the limits on the number of designations, weekly frequencies, charter operations, codeshare arrangements, cargo operations, and other business matters, although the model text of an open skies agreement could not prevail until March 2010. A specific time schedule was put forward for the Chinese side to agree in respect of removing the existing constraints. Specifically, the USA intended to achieve the following:

- With respect to the number of designated carriers, the USA was happy to see an unlimited designation of Chinese carriers with immediate effect and an unlimited designation of US carriers effective as of March 2010. Before that, the US side wanted to have another carrier become eligible for operations starting from the summer season 2008.
- With respect to traffic rights and capacity, the USA wanted to move Guangzhou from Zone 1 to Zone 2 and allow their airlines to operate as many flights as possible to Beijing, Shanghai, and Guangzhou. They also wanted to have unlimited frequencies for cargo operations with immediate effect. To facilitate their operations, the USA asked for the freedom of choosing one more intermediate point at its will for combination carriers.
- With respect to codeshare arrangements, the USA wanted to have more airlines to be allowed to have codeshare arrangements in either country and remove the conditions attached to the third-country codeshare arrangements. For charter operations, the USA wanted to have a certain amount of charter flights to be approved over the next two years with the standard open skies text with respect to charter operations prevailing in 2010. To promote economic growth in Guam and the Northern Mariana Islands, the USA welcomed carriers of both sides to start operations without any restrictions.

March 2010 was targeted for the USA and China to agree a truly open skies agreement with the standard open skies text applicable. This would remove zone division in China with respect to the destinations of operations, with the US carriers being allowed to operate from any points in their home country to any points in China with any frequencies.

Acknowledging such an aggressive proposal from the US side, China was not prepared to conclude anything that would be substantially different from what had been agreed in 2004. In China's view, it was not the right time yet for China to agree such a deal although it had taken a positive approach toward liberalisation in international air transport; the government believed that only a progressive and phased-in manner was appropriate. They were worried about the spill-over effects of such prospective "open skies" arrangements, which might inspire other countries and regions such as Singapore, Japan, South Korea, and the EU to follow suit and demand "open skies" for their carriers (Wolf, 2007). Nevertheless, China accepted the fact that the implementation of the 2004 Protocol had exerted a positive impact on the market. For example, the RPK of Sino-US traffic was up 72 percent in 2005 with Air China experiencing an increase of more than 12 percent in 2005 compared with 2004. The Protocol was considered positive in facilitating the hub strategy of Shanghai airport, which saw 96 weekly frequencies originating from the airport to US destinations and 1.77 times more seats in 2005 than that in 2004.

However, on the other hand, the Chinese authority was also concerned about the imbalanced market conditions. For example, the increase in passenger traffic mainly came from the US side while Chinese travellers to the USA saw little increase over the last ten years. Although around 1.65 million passengers travelled between China and the USA in 2006, a 25 percent increase over 2005, Chinese airlines were unable to capture half of the market. In addition, the Chinese were worried about the inadequate infrastructure capability at big airports such as Beijing and Shanghai as a result of the rapid growth of air transport, which were short of slots to accommodate more traffic.

When taking soundings from the industry stakeholders, contrasting views were expressed again. Airports like Guangzhou wanted to have as many passenger and cargo US bound direct flights as possible, to strengthen its hub status. The newly set up all-cargo operators such as Jade Cargo[8] and Great Wall expressed their strong wish to enter the Sino-US market. They argued that the Sino-US market was their priority for international operations and crucial in their network development. They sought to be designated sooner rather than later so as to allow them to take the opportunity of the continuous market growth.

Contrary to the ambitious start-ups, the Chinese incumbents reiterated their objection to a more liberalised arrangement opposing any kind of increase in the number of designated carriers, frequencies, change of gauge, and relaxed codeshare arrangements. They argued that they failed to move into the market as quickly as their US counterparts because they were short of long-haul aircraft and pilots, which were deployed in lucrative domestic markets. Statistics showed that the total number of long-haul aircraft of all Chinese carriers was only 41, less than 1/10th of the fleet of Continental Airlines which had 488 aircraft. The total number of cargo aircraft of all Chinese carriers was only 1/20th of that of UPS. In addition, Chinese airlines criticised the stringent US visa policy which discouraged Chinese citizens from travelling to the USA. They argued that any relaxation of the restrictions on operations would only benefit the US carriers and leave the Chinese industry in a more difficult, less competitive situation.

With varying expectations from different airlines, the CAAC had to consider how to address the imbalanced growth of the industry, how to satisfy the different stakeholders and how to reach an agreement with the USA while at the same time protecting its own industry. With the above considerations in mind, the Chinese believed that the priority for them would be to review the implementation of the 2004 Protocol and address the imbalance rather than to agree to a significant expansion of any arrangements. It would be acceptable to designate more carriers with some reasonable capacity entitlements but not new frequencies to the incumbents. It would welcome more direct flights to the Central, Western, and Northeastern regions of the country but not much increase to the hub airports in Beijing and Shanghai due to their slot shortages. They were happy to see the removal of restrictions on cargo operations but not on passenger services.

With such different agendas, the first round of discussions held in April 2006 did not bring about much consequence. The following months saw both parties meet up several times in attempts to minimise the gap and differences. The USA acknowledged the pressure and challenges encountered by their Chinese counterpart, while the Chinese appreciated the efforts the USA made in giving their support to the Chinese industry. Both parties hoped that meaningful progress could be made towards the amendment of the 2004 Protocol so as to establish a good foundation to fully liberalise the bilateral aviation market within the foreseeable future. Eventually, at the sixth round of discussions held in April 2007 in China, an agreement, the 2007 Protocol, was reached.

Protocol agreed

The 2007 Protocol was officially signed in July in Seattle during the second Sino-US SED discussions, as part of a package reached between the two countries on a wider range of topics. The Protocol further removed restrictions on the designation of carriers, capacity, and codeshare arrangements. For instance, China could designate unlimited number of carriers to operate passenger services to any points between the two countries as of 1 August 2007 while the

USA could apply the rights as of 2011. Both can operate unlimited cargo services between the two countries via any intermediate and beyond points with full traffic rights. In terms of capacity, up to 160 frequencies per week could be introduced to destinations in Zone One in China in a phased-in manner by 2012, compared with 133 for Zone 1 in 2004s Protocol, and unlimited frequencies to be operated to China's Zones 2 and 3, while cargo capacity could be determined freely by carriers without any restrictions. The Protocol had included Guam and the Northern Mariana Islands as US destinations which could be operated with unlimited frequencies by carriers of both parties. The most significant commitment was that both parties acknowledged that the mutual ultimate objective was the full liberalisation of the bilateral air transport market, hence, agreeing to begin, no later than 25 March 2010, to negotiate such an agreement and timetable.

Reactions of stakeholders and implementation of the Protocol

Despite the outcome falling short of the US' expectations (Schofield, 2007), the USA regarded the deal as a significant step towards a full liberalisation with a timetable being agreed for further negotiations. The agreement would stimulate some $5 billion in new business and enable the US carriers to take advantage of China's booming aviation market which had been growing at 17 percent annually and brought the two countries one step closer to each other (Russell, 2007a). The US Embassy officials in Beijing claimed that the agreement would enable US carriers to stay right ahead of the curve with the anticipated significant increase in demand in both passenger and cargo traffic over the next couple of years, and their dominance over the high-yield traffic on Sino-US routes carrying 59 percent of overall passengers in 2006 (Russell, 2007a).

The US carriers applauded the agreement and responded with significant enthusiasm (Russell, 2007b). The elimination of restrictions on cargo operations allowed unfettered access to China by the US shipping giants such as FedEx and UPS, both of which had previously remarkably expanded their services in China by setting up cargo hubs in Guangzhou and Shanghai respectively. Delta, in its search for new and lucrative international routes following a period of services cuts after the 9/11, considered its prospective Atlanta–Shanghai service the best opportunity to recover from its bankruptcy protection (Russell, 2007b). Northwest was able to launch its long-planned direct services to Guangzhou and increase capacities to Beijing and Shanghai. United won the route licence for its planned Washington–Beijing direct flight, the largest segment in the market that had not been served. The DoT selected United contemplating that the first ever service between the two capitals would offer the opportunity to enhance the political and economic ties between the two countries (Beane, 2007).

Prior to the introduction of the service, connectivity between the two countries was between major gateway airports, but with no links from any Chinese airport to Washington. United argued that connectivity between the two capital cities enhanced trade prospects but was also critical for the advancement of political relations. In this regard, it is testimony to the perceived role for air transport in fostering political interaction and influence in the ever evolving post - normalisation era of US-China relations.

Again, similar to 2004, the Chinese side responded with a mix of welcome and criticism. Jade Cargo successfully secured its route licence to operate all-cargo flights while Hainan Airline was awarded the right for Beijing–Seattle services. The incumbents were still concerned about their competitiveness, with China Eastern's chairman Li Fenghua being quoted as saying that he did not want to see international flights open too fast because the income of all the three major carriers on international routes was generally much lower than that on domestic routes (Russell, 2007b). In only six months, four of the US passenger carriers were able to increase their frequencies to 70 weekly (310 per month) attracting unprecedented traffic between the two countries, while the Chinese passenger carriers only managed to maintain 36 weekly (158 monthly) flights. United itself was providing 155 frequencies in January alone (Russell, 2007b) (Table 4.7).

TABLE 4.7 Schedule of Sino-US passenger frequencies in January 2008 (monthly)

Carrier	Route	Frequencies
American	Chicago–Shanghai	31
Continental	Newark–Beijing	31
United	Washington–Beijing	31
	Chicago–Beijing	31
	Chicago–Shanghai	31
	San Francisco–Beijing	31
	San Francisco–Shanghai	31
Northwest	Detroit–Beijing	31
	Detroit–Shanghai	31
	Honolulu–Guangzhou	31
Air China	Beijing–New York	31
	Beijing–Los Angeles	31
	Beijing–San Francisco	31
China Southern	Guangzhou–Los Angeles	22
China Eastern	Shanghai–New York	12
	Shanghai–Los Angeles	31

Source: 2007 Fourth quarterly and annual highlights, www.tinet.ita.doc.gov.

After the 2007 Protocol, China's economy maintained a strong growth and had surpassed Germany in 2008 to become the world's top exporting nation registering $1.47 billion measured by value of the exports, taking 9.2 percent of the world total, while the USA ranked the third, registering $1.3 billion exports in value terms with 8.1 percent of the world total. The USA imported $273 billion from China, accounting for 19 percent of its total, compared with its $81 billion exports to China, accounting for 6.7 percent of its total (DoT, 2010). However, the rapid growth of bilateral trade between the two countries only led some carriers to increase the capacity moderately, while others such as China Eastern and United had adversely decreased their frequencies in 2010 (Figures 4.16 and 4.17). Statistics from the US DoT shows that none of the mainland Chinese airports was able to capture a place in the league of the top ten destination cities for US outbound air freight, although Shanghai airport took a fourth ranking in terms of inbound freight, with 247 thousand tonnes of cargo shipped in 2008 (Table 4.8) (DoT, 2009).

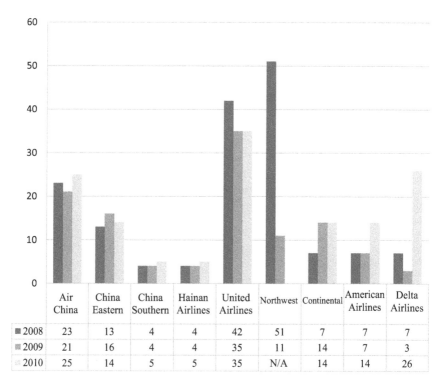

	Air China	China Eastern	China Southern	Hainan Airlines	United Airlines	Northwest	Continental	American Airlines	Delta Airlines
2008	23	13	4	4	42	51	7	7	7
2009	21	16	4	4	35	11	14	7	3
2010	25	14	5	5	35	N/A	14	14	26

FIGURE 4.16 Comparison of weekly frequencies of passenger operations between China and the USA from 2008 to 2010.

Source: www.caac.gov.cn.

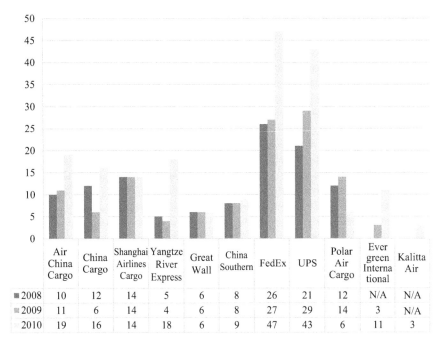

	Air China Cargo	China Cargo	Shanghai Airlines Cargo	Yangtze River Express	Great Wall	China Southern	FedEx	UPS	Polar Air Cargo	Ever green International	Kalitta Air
■ 2008	10	12	14	5	6	8	26	21	12	N/A	N/A
■ 2009	11	6	14	4	6	8	27	29	14	3	N/A
2010	19	16	14	18	6	9	47	43	6	11	3

FIGURE 4.17 Comparison of weekly frequencies of cargo operations between China and the USA from 2008 to 2010.

Source: www.caac.gov.cn.

TABLE 4.8 Top 10 airport pairs for US international outbound and inbound measured by cargo volume: 2008

Rank 2008	US airports	Foreign city	Volume (in thousand tonnes)
Export			
1	Anchorage	Seoul, South Korea	198
2	Anchorage	Tokyo, Japan	157
3	Anchorage	Hong Kong, China	145
4	Anchorage	Taipei, Taiwan China	144
5	Miami	Sao Paulo, Brazil	103
6	Miami	Bogotá, Colombia	93
7	Honolulu	Sydney, Australia	71
8	John F. Kennedy, New York	London, UK	63
9	O'Hare, Chicago	Frankfurt, Germany	49
10	John F. Kennedy, New York	Brussels, Belgium	49

(*Continued*)

TABLE 4.8 Continued

Rank 2008	US airports	Foreign city	Volume (in thousand tonnes)
Imports			
1	Anchorage	Seoul, South Korea	455
2	Anchorage	Taipei, Taiwan China	328
3	Anchorage	Hong Kong, China	316
4	Anchorage	Shanghai, China	247
5	Miami	Bogotá, Colombia	198
6	Anchorage	Tokyo, Japan	154
7	Miami	Guayaquil, Ecuador	89
8	Los Angeles	Tokyo, Japan	88
9	Miami	Lima, Peru	87
10	Miami	Santiago, Chile	76

Source: DoT (2009).

The future

It has been almost four decades since China and the USA resumed bilateral air services. Starting from two designations on two routes with four frequencies in 1980 to unlimited designations on unlimited routes to Zones 2 and 3 with unlimited frequencies in 2007, the arrangements have significantly facilitated bilateral trade and enhanced the exchange between the two countries. Table 4.9 highlights the key points of the Sino–US agreements concluded over the past three decades, while Table 4.10 provides detailed information of the airlines operating in the market as of 1980.

Over the years, the USA has been proactively promoting "open skies" around the world with a view that airlines are entitled to engage in business and international trade without restrictions on competition. "Freedom of the air" should facilitate international trade and tourism by acknowledging the rights of airlines to participate in the air transport industry on the basis of equal opportunity (Beane, 2007). On the other hand, China believes that it has made significant progress in opening up its market and integrated into the global economic system, and that international air transport is a political instrument as well as an economic activity which should be used to leverage and optimise the national interests (Li, 2010). With a "proactive, progressive, orderly and safeguarded" approach towards liberalisation of international air transport, China holds that the pace for liberalisation should be appropriate to allow Chinese airlines to enhance their competitiveness and develop the capability of dealing with the open skies circumstances effectively (Li, 2010). Only in so doing, can Chinese airlines not be forced out of the market as a result of open skies due to their weak marketing, poor branding, and shortage of management skills (Wolf, 2007).

TABLE 4.9 Key points of bilateral arrangements over the past three decades

Year	Number of designated carriers	Routes	Capacity (total weekly frequency)	Fifth freedom traffic rights	Codeshares
1980	2 for each side	2	4	No	No
1992	3 for each side	3	27		Yes
1999	4 for each side	3	54	No limitation at intermediate and beyond points	Yes
2004	9 for each side	Unlimited	133 between any points in the US and Beijing, Shanghai and Guangzhou by April 2007; 249 in total for all destinations	No limitation on cargo, Limited for passengers between China and Japan sector	Yes
2007	Unlimited for China, 11 for the US	Unlimited	227 between any points in the US and Beijing, Shanghai and Guangzhou by April 2010;	No limitation on cargo, Limited for passengers between China and Japan sector	Yes

Source: Adapted from the amendments concluded in 1980, 1992, 1999, 2004, and 2007.

TABLE 4.10 Carriers operating in Sino-US market with year of market entry

	Combination service		All cargo service	
Year	Chinese carrier	US carrier	Chinese carrier	US carrier
1980		Pam Am		
1981	CAAC			
1984		Northwest		
1986		United replace PamAm		
1987	Air China to replace CAAC			
1991	China Eastern			
1992				Evergreen International
1993			Air China Cargo	FedEx replace Evergreen International

(Continued)

TABLE 4.10 Continued

	Combination service		All cargo service	
Year	Chinese carrier	US carrier	Chinese carrier	US carrier
1997	China Southern		China Cargo replace China Eastern's cargo operation	
2001				UPS
2004			Shanghai Cargo	Polar Air Cargo
2006		Continental American		
2007			Yangtze River Express	
2008	Hainan Airlines	Delta	Great Wall Jade Cargo	
2009				Evergreen International
2010				Kalitta Air

Source: Compiled by the authors from various sources.

The increasingly integrated trade relationship between the two countries will no doubt require more transportation to satisfy business needs. Despite the commitment made in the 2007 Protocol that both parties would resume discussions no later than March 2010, the two actually did not meet up until June with little being achieved due to divergent interpretations about the timetable for a full open skies deal, as well as different considerations that both parties have. As liberalisation has now become an irreversible trend for international air transport, China has to keep up with the pace. Looking into the next round of discussions, the US side was hopeful of determining a roadmap for a full open skies within a certain period of time, while the Chinese, though acknowledging the commitment of agreeing to the deal, wanted a longer period of time so as to get its industry well prepared. As both parties were in close contact in negotiating the deal, it still remains to be seen when full liberalisation will be achieved. The ongoing trade war and strained trading relations even further complicate the extent to which a speedy resolution can be reached with no real progress having been made since the 2007 consultation.

Notes

1 CNTA was merged into Ministry of Culture, which became Ministry of Culture and Tourism in 2018.
2 In 1985, Chinese President Li Xian'nian visited the USA, which was the first official visit of a Chinese President to the USA.
3 China Southern first entered the Sino-US market in 1997.
4 FedEx replaced Evergreen International in 1993.
5 China has divided its aviation market into three zones: Zone One includes the airports of Beijing, Shanghai, and Guangzhou; Zone Two includes airports in Eastern and coastal cities such as Nanjing and Hangzhou; while Zone Three refers to the rest

of the provinces in Central, Western, and Northeastern China such as Henan, Anhui, Hubei, Hunan, Jiangxi, Guangxi, Guizhou, Yunnan. Sichuan, Chongqing, Shaanxi, Shanxi, Gansu, Ningxia, Qinghai, Tibet, Xinjing, Heilongjiang, Inner Mongolia, Jilin, Liaoning, and Hainan Island. The 249 flights involved cities are in Zones One and Two only. For Zone Three, there is no limitation for any operations.
6   Except American Airlines which did not launch their service until 2006.
7   Air China reduced one weekly frequency on the Beijing–San Francisco route.
8   Jade Cargo received the US DoT approval for an all-cargo charter operation in March 2007. The flight was changed to a scheduled operation after the 2007 agreement.

References

ATA, (2010), When America flies, it works, 2010 Economic Report, available at www.airlines.org/Economics/ReviewOutlook/Pages/AnnualEconomicreportsoftheU-SAirlineIndustry.aspx

BBC, (1999), 1999: Chinese anger at embassy bombing, available at http://news.bbc.co.uk/onthisday/hi/dates/stories/may/9/newsid_2519000/2519271.stm

Beane A. R., (2007), Aviation relations between the United States and China: are open skies on the horizon? *Air Law and Commerce*, 72, 803.

Beningo S., (2008), US-China trade growth ad America's transportation system, April 2008, Bureau of Transportation Statics, Special Report, SR-007US Department of Transportation, Research and Innovative Technology Administration.

Bureau of Transportation, (2018), Airport Rankings 2018, available at https://www.bts.gov/airport-rankings-2018

CAAC, (2005), *Statistical data on Civil Aviation of China, 2005*, Civil Aviation Press of China, Beijing.

Caijing Daily, (2004), Sino-US Air Pact Causes Domestic Panic, 05 July, 2004, available at www.english.caijing.com.cn

Chen N., and Lee J. L., (1984), Reagan visit to China, accords support expansion of US export, trade recovers from slump, could reach new high, 20 August 1984, Business America, available at http://findarticles.com/p/articles/mi_m1052/is_v7/ai_3395321/

Clancy M., (2007), Globalizing the skies: Domestic and international sources of the liberalisation of commercial air transport, Paper for 2002 International Studies Association Meeting in New Orleans.

Compart A., (2008), Spanning the globe: will adding international service save US airlines? *Travel weekly*, available at http://travelweekly.com

Deng X., (1990), The international situation and economic problems, The selected works of Deng Xiaoping, Vol. 3 (online version), available at http://english.peopledaily.com.cm/dengxp/vol3/text/d1130.html

DOT, (2018), 2017 traffic data for U.S airlines and foreign airlines U.S. flights. available at www.bts.gov

DOT, (2010), 2009 traffic data for U.S airlines and foreign airlines U.S. flights. available at www.bts.gov

DOT, (2009), 2008 traffic data for U.S airlines and foreign airlines U.S. flights. available at www.bts.gov

Feng H., (2006), *The politics of China's accession to the World Trade Organization*, Routledge, London and New York.

Garrison J., (2005), Understanding prospects for cooperation and competition in US-China relations–a review of theory and practice. Paper prepared for the annual meeting of the International Studies Association, Honolulu, Hawaii, USA.

Havel B. F., (2009), *Beyond open skies, a new regime for international aviation*, Kluwer Law International, The Netherlands.

He K., (2009), Dynamic balancing: China's balancing strategies towards the United States, 1949–2005. *Journal of Contemporary China*, 18(58), 113–136.

Head M., (1999), How could the bombing of the Chinese embassy have been a mistake? 10 May 1999, available at www.wsws.org

Hong Kong Trade Development Council (2004), US, China Ink Landmark Air Service Agreement, 11 Aug 2004, available at www.hktdc.com

IATA Press Release No.4, 01 February, (2010), Asia Pacific: challenges and opportunities – Intra-Asia Market Eclipses North America as World's Largest, available at www.iata.org/pressroom/pr/Pages/2010-02-01-01.aspx

Khalilzad Z., (1999), *The United States and a rising China: strategic and military implications*, Washington, RAND, available at www.rand.org

Lampton D. M., (2001), *Same bed, different dreams: managing US-China relations, 1989–2000*, University of California Press, Berkeley and Los Angeles.

Li J. X., (2010), Calling on all industry to strive for a strong air transport country of China, 14th February, 2010, available at www.caac.gov.cn/D1/MHQG/201007/t20100719_33631.html

Li W., (1994), The impact of economic reform on the performance of Chinese state enterprises 1980–1989, Duke University, available at http://129.3.20.41/eps/dev/papers/9410/9410001.pdf

Lim C., and Wang. Y., (2008), China's post-1978 experience in outbound tourism. *Mathematics and Computers in Simulation*, 78 (2008), 450–458.

Lynch D. J. (2005), US-Chinese trade relations get trickier, 13 September, 2005, USA Today, www.usatoday.com/money

Meyer G. S., (September 2002), US-China aviation relations: flight path toward open skies? *Cornell International Law Journal*, 35, 427.

Oxford Economics (2016), The importance of air transport to the United States, available at www.iata.org/policy/Documents/benefits-of-aviation-usa-2017.pdf

Payload Asia, (2007), US, China aim for new open skies pact, Payload Asia, May 2007, available at www.payloadasia.com/article-6-chinauschinaaimfornewopenskiespact-PayloadAsia.html

Russell E., (2007a), China, US expand air traffic, 31 May 2007, available at www.atimes.com

Russell E., (2007b), Six US-China air routes cleared for takeoff, 29 September, 2007, available at www.atimes.com

Schofield A., (2007), U.S.-China Open-Skies Deal Remains Elusive, 01 May 2007, available at www.aviationweek.com

Shi T., with Wen M. C., (2009), Avoiding mutual misunderstanding: Sino-US relations and the new administration, January 2009, Carnegie Endowment for International Peace, Foreign Policy Summary.

Steinberg A. B., (2007), The importance of aviation liberalization, Speech delivered at the US-China Business Council Aviation Forum, January, 2007, Beijing, China.

Sweeney J., Holsoe J., and Vulliamy E., (1999), Nato bombed Chinese deliberately, 17 October 1999, available at www.guardian.co.uk

Tanger R. H., (2007), *The air cargo market between China and the United States: demand, developments and competition*. Kellogg School of Management Northwestern University, Evanston, IL.

The US Department of Commerce, International Trade Administration, Office of Travel & Tourism Industries, (2007), International Arrivals to the United States, Fourth quarter and annual highlights; US, available at www.tinet.ita.doc.gov

The US Department of Commerce, International Trade Administration, National Travel and Tourism Office (2017), International Arrivals to the United States, Market Profile of Overseas Visitor; US, available at https://travel.trade.gov/

The US Department of State, (2010), open skies partners, available at www.state.gov/open-skies-partners/

US DOT (2020), About us, available at: https://www.transportation.gov/about

Wang Y., (2002), Facts and figures of Sino-US Trade in 2001, 27th February, 2002, available at www.china.org.cn

Wolf D., (2007), Open skies agreement between U.S. and China? not anytime soon, 19th April 2007, available at www.seakingalpha.com

Yamanouchi K., (2010), Delta eyes Asia growth, return to Shanghai, The Atlanta Journal-Constitution, December 2010, available at www.ajc.com/business/delta-eyes-asia-growth-768670.html

Zeng K.,(2004), *Trade threats, trade wars: bargaining, retaliation, and American coercive diplomacy*, The University Press of Michigan, Ann Arbor.

Zhang A., and Chen H., (Spring 2003), Evolution of China's air transport development and policy towards international liberalisation. *Transportation Journal*, 42(3), 31.

Zhang W., (2008), Background information: milestones of Sino-US relations, 31st December, 2008, available at www.xinhuanet.com

5

CHINA, THE NETHERLANDS, AND THE UK

China and the Netherlands

The Netherlands and its air transport industry

Comprising only 34,000 square kilometres with more than 16 million people, the Netherlands is strategically situated right at the heart of Western Europe, which provides a main gateway to Europe and the northwest of the continent, making it one of the largest and most important transport and distribution centres in the world. The opening of the EU's internal borders and the introduction of the euro have strengthened the Netherlands' leading position in transportation, distribution, and logistics, enhancing the competitive strength of its transport industry which accounts for 8 percent of the country's GDP and a large share of the European market.

As a shift in economic activities is occurring around the world, the government fully appreciates the demand for transport and logistics and is ambitious to reinforce the country's role as the key European hub. The White Paper for Aviation issued in April, 2009 recognised that the aviation sector would play a pivotal role in the country's aim to further enhance its allure and appeal for business and to remain as one of the most competitive economies of the world. Schiphol Airport and KLM were regarded as playing a key role in the country's initiative. The government expressed its commitment to encouraging Schiphol Airport, one of the home bases and hubs for Air France–KLM and its SkyTeam partners, together with other national airports to develop an extensive route network so as to offer the most convenient connections appropriate to the desired spatial–economic development of the region and to link the country directly to the world's major established and new economic centres (The Ministry of Infrastructure and Environment, 2009). The White Paper consequently informed the development of the National Airspace Policy.

Amsterdam Schiphol Airport (AMS) was ranked Europe's third largest airport in terms of passenger throughputs in 2018, handling 499,444 air transport movements and 71 million passengers, just behind Paris Charles de Gaulle (72.2 million) and London Heathrow (80.1 million) (Schiphol Airport Annual Report, 2009), representing 11.7 percent of market share of European traffic. In terms of cargo, the airport recorded declines in its cargo volumes, handling 1.76 million tonnes of freight declining by 2.5 percent from 1.71 million tonnes in 2017 (Schiphol Group 2019). Of this, 56 percent was carried via full freight operations with 44 percent on passenger flights. The decline was said to be attributed to the significant losses in slots by the full freight carriers.

Notwithstanding, Schiphol continues to be an important European gateway for China both in terms of passenger and cargo traffic. In 2018, it served as the best-connected airport for Chinese travellers carrying 1.8 million to and from Europe and China. Direct services have been established to seven Chinese cities, namely, Beijing, Chengdu, Hangzhou, Guangzhou, Shanghai, Xiamen, and Hong Kong (SAR). Chinese airlines servicing the country pair included China Southern with seven flights per week from Guangzhou and Beijing, China Eastern with four flights per week from Shanghai, Xiamen Airline with a thrice weekly service from Xiamen, and Cathay Pacific with a daily service from Hong Kong. KLM Royal Dutch Airlines offered the most prominent service with 12 flights per week to Shanghai amid other services to Beijing, Hangzhou, Chengdu, and Hong Kong (SAR). China–EU trade also creates a market for EU products in China with substantive interests being directed towards products from the Netherlands. Schiphol airport estimates that cargo between the Netherlands and China amounted to 431,000 tonnes in 2018.

With Schiphol holding such a critical position in the China–European air corridors, it is also important that growth at AMS is not restricted by capacity constraints set forth by the Alders agreement which established a limit of 500,000 air traffic movements until 2020. The agreement was formed between the airport, and various stakeholders including state representatives, local and regional authorities, airport parties, and municipalities to balance the interest of the airport and that of other stakeholders. The airport nonetheless continues to maintain its position as a major European hub with only 33 percent of passengers being residents of the Netherlands. Compared with 2008 where AMS ranked sixth in its direct connectivity, the airport ranked first in 2017 and second in 2018 having been overtaken by Frankfurt.

KLM is the world's oldest airline still operating under its original name. Set up in 1919 but merged with Air France in 2004 to form the Air France–KLM Group and as the founding members of SkyTeam, it became the world's largest airline partnership in terms of financial turnover by 2009. Operating under its own brand with a fleet of 214 aircraft, KLM served 166 destinations in 63 countries and achieved 107 billion RPK and 4.8 billion RTK in 2018. In 2018, the group generated revenues totalling 11 billion euros compared with the 10.5 billion in 2017.

The government agency responsible for international air transport policies

Responsibility for the country's transport system including aviation was held by the Ministry of Transport, Public Works, and Water Management. The Aviation Negotiation Office headed by the chief negotiator was the unit mainly overseeing the development of the nation's international air transport policies and negotiating bilateral air services with its counterparts. To optimise the resources and better coordinate work, the Ministry of Transport, Public Works, and Water Management was subsequently merged with the Ministry of Housing, Spatial Planning, and the Environment to form the new Ministry of Infrastructure and Environment. The Aviation Office retained its responsibilities reporting to one of its Directorate Generals (Directorate General for Mobility and Transport). The administrative disposition was intended to develop and maintain strong transport connections through multi-mode provisions to improve traffic flow with an aim of strengthening its strategic position as a hub for Western Europe and further into Africa, Asia, and Eastern Europe.

The views towards an open economy

In spite of its small size, the Netherlands' wide receptiveness to international trade and investment has helped to foster a relatively large economy. Over the centuries, the country has featured an open and export-oriented economy with trade and distribution being one of the important driving forces (Staniland, 2003). In light of this openness to global trade, and positioning itself as a model in global commercial success, the government of the Netherlands would obtain traffic rights where possible to support rapid international expansion. Like a cohesive system, Dutch airlines would concurrently strive towards fiscal and operational prudence to ensure they are competitive and efficient to sustain their profitability.

Such an open view towards the economy has enabled the government and its businesses to be proactive and innovative in pursuing market opportunities. Removing restrictions for a liberalised operational environment has been one of the main objectives of the Netherlands government in supporting its businesses in their international commercial expansion initiatives. Such a philosophy has been guiding the Netherlands government in its bilateral air transport negotiations with the rest of the world, including the USA and China. It was the first country that signed the early form of an open skies agreement with the USA in 1978, which gave KLM access to major US cities such as Los Angeles and Atlanta. It was again the first country that signed a real open skies agreement with the US in 1992 that built on the strategic alliance formed between KLM and Northwest Airlines in 1988, which not only led the way for global airline alliances but also marked a new era for the global aviation industry.

Bilateral air transport negotiations between China and the Netherlands

Although China and the Netherlands started their diplomatic relations in 1954, the bilateral ASA had not been signed and did not become effective until May 1996. Ever since then, four rounds of consultations have been conducted, respectively, in 1998, 2003, 2006, and 2010, which altogether have removed the restrictions on airline operations between the two countries, albeit the process being gradual and progressive. In 2010, the Netherlands became one of the first EU member states that amended its bilateral ASA with China to bring it into conformity with the EU law, allowing EU designation. The following is an overview of the bilateral consultations as they evolved.

Negotiating the bilateral ASA

The diplomatic relationship that began in 1954 between China and the Netherlands did not bring about an immediate discussion and conclusion of a bilateral ASA. Rather, it took both parties 25 years before they sat together for such a negotiation, resulting in the conclusion of an ASA in January 1979. However, the agreement failed to enter into force due to the Chinese side failing to complete the legal procedures as a consequence of the air transport services launched by KLM to Taiwan (part of PR China, hereinafter referred to as Taiwan) in January 1983. The service was considered a serious political offence by China, which urged the Netherlands to address the issue to its satisfaction. The following years saw both sides exchange views and visits of senior officials, with an intention to resolve the issue and remove the legal obstacles preventing the launch of direct services between the two countries. The efforts eventually led to a satisfactory resolution in May 1995 when both sides reached an understanding regarding the Netherlands–Taiwan service. The Netherlands agreed that the Taiwan service would be treated as a private commercial operation and China Airlines of Taiwan would remove its national flag on all its flights as of October 1995.

In July 1995, discussions were resumed between the two parties leading to the agreement of a fresh ASA, which was subsequently signed in May 1996 in Beijing and became valid with immediate effect. The agreement allowed both sides to designate one carrier from each side to operate between the two countries with two intermediate points to be discussed and agreed (Table 5.1).

The Netherlands designated KLM as its carrier, which inaugurated a direct Amsterdam–Beijing service with two weekly frequencies in April 1996. For the Chinese side, the CAAC had to decide which of the three bidders, that is, Air China, China Eastern, and China Southern, was more appropriate for this new operation. Although Air China operated to quite a number of European destinations and China Eastern had just launched China–Germany services, neither of them was successful in this bidding. Instead, China Southern, which was very new to the European market, was awarded the traffic right, which

TABLE 5.1 Arrangements agreed in 1996 between China and the Netherlands

Items	Chinese side	The Netherlands side
Routes for passenger and cargo and weekly capacity	2 frequencies for China to Amsterdam	2 frequencies for Amsterdam to Beijing for passenger service; To be determined by the designated cargo carrier, operating on a charter basis, between Amsterdam and Guangzhou
Number of designated carriers	1	1 for passenger 1 for cargo charter operator

Source: Adapted from the Sino-Netherlands ASA agreement in 1996.

argued that it needed long-haul international services to expand its Southeast Asia-focused network. The carrier launched its first-ever European service between Guangzhou and Amsterdam via Beijing in November the same year. At the same time, it cooperated with Martinair on the carrier's charter cargo service between Amsterdam and Guangzhou.

Cautious expansion of traffic rights in 1998

The initial operation of services generated quite some market interest though the demand remained moderately low (Rowland, 2010). KLM was able to achieve 73 percent load factor on its Amsterdam–Beijing route compared with the 42 percent load factor of China Southern on its Guangzhou–Beijing–Amsterdam service in early 1998, two years following the launch. A further consultation was scheduled in June 1998, with one of the objectives from the Netherlands side being to have more frequencies and destinations in China (Rowland, 2010). Martinair, the cargo operator that had been flying to Guangzhou on a charter basis, also wanted to become a designated schedule operator.

The eagerness of the Netherlands side to expand operations in the market was not echoed by China. China Southern was not enthusiastic about increasing capacity, nor did it want to see more competitors in the market. It argued that its extensive domestic network and focus on Southeast Asia failed to generate sufficient demand for the China–Netherlands market, which was their only route to Europe. In addition, it was not making any money from its international operations due to the financial crisis in Southeast Asia in 1997. It was concerned that any increase of designated carriers and capacity would only adversely impact the carrier's ability to compete and operate profitably. It therefore suggested that limitation should be put in place to restrict more carriers from entering into the market so that the carrier could have more time to develop the market.

China Eastern also expressed the view of not including Shanghai as an additional destination in China for the Netherlands, arguing that the airport was short of facilities to accommodate more operations. The fierce competition in

TABLE 5.2 Arrangements between the two countries in 1998

Items	Chinese side	The Netherlands side
Routes for passenger and cargo and weekly capacity	7 in total for China to Amsterdam	7 in total, 3 for Amsterdam to Beijing 2 for Amsterdam to Shanghai 2 for cargo services from Amsterdam to points such as Shanghai, Nanjing, and Guangzhou
Number of designated carriers	2	2

Source: Adapted from the Sino-Netherlands ASA amendments in 1998.

the Shanghai market had caused the dramatic decrease of load factors on its services, thus, a protective mechanism should be in place for the benefit of the Chinese airline industry.

Despite the divergent views of the relevant carriers, a package was agreed between the two authorities which moderately relaxed the restrictions on operations for both sides. Routes for passenger and cargo traffic were separated to allow more carriers' access to the market. Shanghai was agreed to be another destination for the Netherlands, although operations would not start until 1999 when Pudong airport would be open. Frequency increase was allowed to Beijing, and both intermediate and beyond points were agreed for cargo operation along with full fifth-freedom traffic rights (Table 5.2).

The arrangements fulfilled the requirements of Martinair as a second designated carrier to operate an all-cargo service between Amsterdam and Guangzhou provided that Martinair would continue collaboration with China Southern on its cargo operations. It also gave KLM access to the Shanghai market following other European carriers such as Lufthansa, Air France, and SAS, and satisfied its request to increase frequencies on Amsterdam–Beijing to three weekly. Nanjing, a city 254 kilometres east to Shanghai was also included as a destination for cargo operations.

The 2003 agreement

The years following 1998 were a period of honeymoon for China and the Netherlands, both of which enjoyed bourgeoning economic exchanges. Between 1998 and 2003, the bilateral trade between the two countries had grown remarkably, reaching a historical high in 2003 with a total trade value of €16.4 billion compared with €5.4 billion in 1999, rendering the Netherlands the second largest trade partner of China in the European Union (Eurostat, 2004).

The traffic demand reflected the sudden booming of the bilateral trade which led the carriers to request an increase in capacity. Both KLM and China Southern applied to their respective authorities to fly more frequencies (Daniels, 2010; Rowland, 2010). They also agreed on codeshare arrangements on their operations covering beyond points which circumvented the regulatory restrictions imposed in the previous arrangements in 1998. Martinair, the cargo operator, wanted to utilise the fifth freedom traffic rights to enable operations via Nanjing to destinations beyond China.

Their efforts were appreciated by their respective authorities. By 2003, before the official discussions were resumed, both KLM and China Southern had successfully managed to have their frequencies and capacity increased to satisfy the market demand. For example, KLM was operating four frequencies on its Amsterdam–Beijing route and five on the Amsterdam–Shanghai route, while China Southern was operating four frequencies on its Guangzhou–Beijing–Amsterdam route (two more than the provisions in the 1998 arrangements). Martinair was able to use the fifth freedom rights granted to extend its services to Bangkok and Sydney via Nanjing.

With such encouraging and promising performance of both sides, the discussions held in 2003 were very pleasant, with the arrangements welcomed by both sides. Compared with the previous agreement in 1998, restrictions on the number of origin airports were removed, allowing carriers of both sides to originate their services from any cities within their respective territories. For passenger services, a total of 25 frequencies were agreed for each side with any type of aircraft, representing a 400 percent increase compared with that in 1998 (five frequencies in total). Specifically, KLM could fly seven weekly frequencies on the Amsterdam–Beijing route and ten frequencies on Amsterdam–Shanghai route, with the rest being allocated to destinations in China such as Guangzhou and Chengdu. Chinese carriers could operate up to 25 weekly frequencies on any routing between China and the Netherlands. On the Amsterdam–Shanghai route, both passenger and all-cargo services were allowed. For cargo operations, both sides were given the traffic rights for 14 weekly frequencies. Codeshare arrangements were further encouraged to promote closer collaboration. Table 5.3 summarises the key points of the agreement reached in 2003.

TABLE 5.3 Arrangements reached for passenger operations in 2003

Items	Chinese side	The Netherlands side
Passenger routes and weekly capacity	25 in total for China to the Netherlands and beyond points	7 for Amsterdam to Beijing 10 for Amsterdam to Shanghai 8 for Amsterdam to other Chinese destinations
Cargo route and weekly capacity	14 in total for China to the Netherlands and beyond points	14 in total for Netherlands to China and beyond points
Number of designated carriers	1 for passenger services 1 for cargo	1 for passenger services 1 for cargo

Source: Adapted from the 2003 ASA amendments.

Arrangements reached in 2006

The agreement reached in 2003 had significantly enlarged the traffic rights between the two countries, although restrictions were still in place limiting the number of designated carriers and frequencies on both passenger and cargo operations. Although there was no evidence that the market would require such a big increase in capacity, both governments were visionary and generous in agreeing to such an arrangement, which enabled the carriers of both sides to be well prepared for the prospect of rising market demand.

Since 2003, the bilateral trade between the two countries has kept momentum, with the Netherlands remaining as China's second largest trade partner among the EU member states. In 2005, the total trade value reached $28.8 billion, a 34 percent increase year on year, making China the fourth largest trade partner for the Netherlands. By 2006, hundreds of businesses from the Netherlands had made a prominent presence in the Chinese market including big brands, such as Phillips, Shell, and Unilever, indicating a promising trade prospect between the two countries.

Corresponding with the surge of bilateral trade, both passenger and cargo traffic continued to grow rapidly, with a total of 427,500 passengers and 118,708 tonnes of freight being carried between the two countries in 2005 (CAAC, 2005). As a consequence, more carriers demonstrated significant enthusiasm in the market, with two Chinese all-cargo operators, that is, Jade Cargo and Great Wall launching an all-cargo service in 2006 under provisional route licences due to the restrictions imposed in the 2003 agreement, thus bringing the total number of carriers to five.

Notwithstanding the phenomenal traffic growth between the two countries, airlines had not used up all of their traffic rights by the winter/spring season 2006 (CAAC, 2006). Tables 5.4 and 5.5 show the status of the utilisation of traffic rights of carriers of both sides, which reveals that although both KLM and China Southern had used their seven frequency entitlement on Amsterdam–Beijing passenger route, KLM had only used seven out of its ten frequency entitlement on Amsterdam–Shanghai passenger route. In terms of cargo services, neither side had used up their 14 weekly frequency entitlement.[1]

TABLE 5.4 Status of utilisation of traffic rights in 2006 for passenger services

Airlines	25 frequency entitlement for passenger operations for each side	Actual utilisation of the weekly frequencies (in percentage)
KLM	Amsterdam–Beijing, 7 Amsterdam–Shanghai, 7 Amsterdam–Chengdu, 2	16, (55% of the total)
China Southern	Guangzhou–Beijing–Amsterdam, 7	7, (24% of the total)

Source: www.caac.gov.cn.

TABLE 5.5 Status of utilisation of traffic rights in 2006 for cargo services

Airlines	14 frequency entitlements for cargo operations for each side	Actual utilisation of the weekly frequencies (in percentage)
KLM	Amsterdam–Shanghai, 4	9 (64% of the total)
Martinair	Amsterdam–Shahjar–Tianjin–Nanjing– Bangkok, 5	
China Southern	Shenzhen–Shanghai-Amsterdam, 5	8 (57% of the total)
Jade Cargo	Shenzhen–Amsterdam, 3	

Source: www.caac.gov.cn.

In this circumstance, passenger carriers were keen to increase their capacity on the Beijing and Shanghai routes, while cargo operators longed for being officially designated. In addition, more cargo operators from the Netherlands side wanted to be designated while Chinese airports such as Nanjing and Xiamen would like to see more fights with full fifth freedom traffic rights.

With no mounting pressure from either government, nor any special requests from airlines, airports, or other stakeholders, discussions in November 2006 were held in a very delightful environment, with an arrangement agreed which further removed restrictions on market entry and capacity. Briefly, the total number of designated carriers for both sides was to be increased to five, which could thus accommodate the requirements of Chinese cargo operators such as Jade Cargo and Great Wall. For passenger services, carriers of both sides were allowed to increase to a total of 33 frequencies from the previous 25. For Netherlands carriers, the maximum frequency for Amsterdam–Beijing route was nine flights per week while Amsterdam–Shanghai was increased to 14 flights per week. For flights to other cities in China a maximum of seven weekly flights was permitted. For cargo operations, a total of 25 frequencies were allowed for both sides compared with the previous 14, but the Netherlands side could not increase their frequency to seven on the Shanghai cargo route until 2008. The agreement also allowed fifth freedom traffic rights and codeshare arrangements between the designated carriers. Table 5.6 provides details of the agreements.

Compared with the agreement reached in 2003, the market entry restrictions were only relaxed modestly, with the Netherlands side being given more destination points in China, while the Chinese side being allowed to introduce two more cargo operators into the market. Passenger capacities were increased with two more frequencies to Beijing, and four more to Shanghai. For cargo operations, capacity was allowed almost 50 percent increase for each side, with fifth freedom traffic rights being permitted at immediate and beyond points. Codeshare arrangements were encouraged with third country carriers to enable the airlines to expand their network.

TABLE 5.6 Agreement reached in November 2006 for passenger and cargo operations

Items	The Netherlands side	Chinese side
Number of designated carrier	5	5
Routes for passenger and weekly capacity	33 in total: 9 for Amsterdam–Beijing (phased in until 2008) 14 for Amsterdam–Shanghai (phased in until 2009) 10 in total for Amsterdam to other cities including Chengdu and Guangzhou	33 for Points in China and Amsterdam
Routes for cargo and weekly capacity	25 in total Amsterdam to points including Shanghai, Nanjing, Shenzhen, and Guangzhou; maximum 7 to each point	25 for points in China and Amsterdam

Source: Adapted from the 2006 ASA amendments.

Consultation and arrangements reached in March 2010

Since 2006, bilateral relations between the two countries have been further strengthened, with more frequent senior level visits and continuous strong growth of trade. According to China's MoC, in 2006, the trade between the two countries totalled $42.7 billion, with China exporting $38.8 billion worth of products. In 2008, the figure jumped to $64.7 billion with $58.9 billion of exports (Figure 5.1), with the Netherlands remaining the second largest trade partner for China in the EU, although the financial crisis had an impact on bilateral trade, causing a sharp drop to $41.8 billion in 2009, a 18.4 percent decrease. In 2009, the Netherlands became the third largest source of investment recording $790 million in China among the EU member states in 2009, compared with $865 million in 2006[2] (Wong, 2009).

The ADS of EU countries for Chinese tourists in 2003 stimulated China's outbound travel market to Europe, resulting in a remarkable rise in Chinese tourists to Europe. At the same time, arrivals from Europe including the Netherlands to China increased steadily. In 2005 around 4.78 million Europeans visited China, among whom, two million arrived by air. In 2008, the figure rose to 6.1 million, with 2.48 million chose to fly to China, though the figure dropped to 4.6 million in 2009 due to the economic downturn. Figure 5.2 shows the arrivals of air passengers in China from the selected EU member states compared with the USA between 2005 and 2009.

The strong demand in both passenger and cargo operation after 2006 inspired airlines to invest in the market to satisfy the demand. Soon after the 2006 consultations, both KLM and China Southern announced an increase of frequencies. In the winter/spring season of 2008, KLM was serving Beijing, Shanghai, and

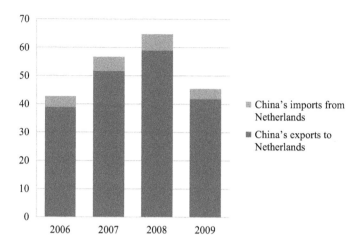

FIGURE 5.1 China's trade value with the Netherlands between 2006 and 2009 (in billions of US dollars).

Source: Adapted from Wong (2009).

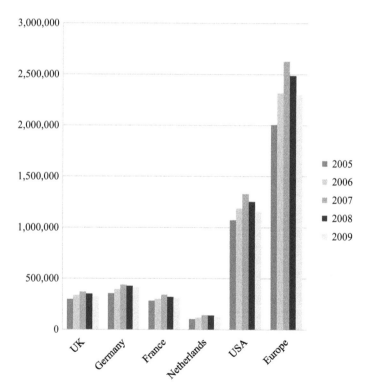

FIGURE 5.2 Arrivals of air passengers from EU and selected member states in China compared with the USA between 2005 and 2009.

Source: www.cnta.gov.cn.

Chengdu with 23 flights per week in total and codesharing with China Southern on 25 routes and Sichuan Airline on six routes (Rowland, 2010). "KLM was proud of being able to provide extensive services with such an extensive network within China with the support from its Chinese partners" (Shanghai Business Review, 2009).

To effectively promote its services, KLM launched a branding campaign in 2008 which reflected its long-term strategy to generate more prospective high-level customers with initiatives including Chinese-speaking crew on aboard the flights, special Chinese menus, onboard Chinese-language magazines, and Chinese-speaking staff at Amsterdam airport to help with Chinese-speaking passengers. In particular, Club China has been launched for those who do business in China to share their experiences (Shanghai Business Review, 2009).

Compared to its counterpart, China Southern in 2007 operated 629 flights carrying 221,308 passengers with an average load factor of 72 percent. Although the number of flights and passengers dropped in 2008 and 2009 due to the effect of the financial crisis on travel demand and international trade, the overall market demand has remained high. Figure 5.3 compares the operational results of the two carriers.

In terms of cargo, the two Chinese carriers, that is, Jade Cargo and Great Wall, acted more efficiently. For new start-ups focusing on the all-cargo international market, high-demand routes imply viable operations, robust business, and profitability. The China–Netherlands market just fit best into the strategy of the two carriers. In 2007, Great Wall operated 11 weekly frequencies carrying

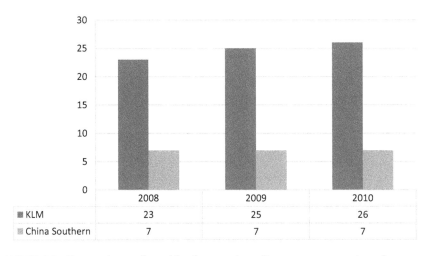

	2008	2009	2010
■ KLM	23	25	26
▨ China Southern	7	7	7

FIGURE 5.3 Comparison of weekly frequencies of passenger operations between China and the Netherlands from 2008 to 2010.

Source: www.caac.gov.cn.

52,000 tonnes of freight while Jade Cargo operated three weekly frequencies carrying 25,000 tonnes of cargo (CAAC, 2007). Although the economic downturn affected their operations, Jade Cargo still managed to fly nine weekly frequencies in 2009. Cargo volume had also been improved due to the utilisation of the fifth freedom traffic rights, which allowed the carriers to take traffic at intermediate/beyond points such as Dubai, Seoul, and Stockholm. By 2010, Jade Cargo had increased its weekly frequency to 19 (compared with its initial three frequencies in 2007), ranking itself the third largest carrier in cargo operations in Amsterdam airport (Sweijen and Pouwels, 2010). Martinair was also able to fly to more Chinese secondary airports, which helped the carrier expand its cargo network in the country. Figure 5.4 compares the performance of all-cargo carriers in the Sino-Netherlands market from 2008 to 2010.

The continuous surge in market demand as well as the permission of the full fifth freedom traffic rights on intermediate and beyond points has attracted more potential Chinese carriers into the market. China Eastern wanted to be a designated carrier after joining SkyTeam while Yangtze River Express, the cargo arm of Hainan Airline group, also expressed its wish for prospective operations

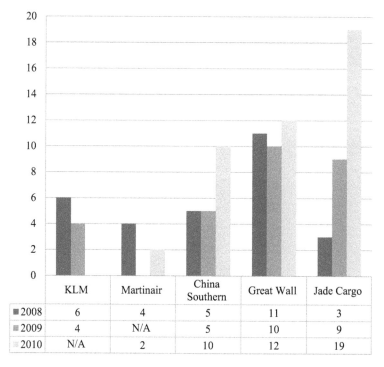

	KLM	Martinair	China Southern	Great Wall	Jade Cargo
■ 2008	6	4	5	11	3
■ 2009	4	N/A	5	10	9
2010	N/A	2	10	12	19

FIGURE 5.4 Comparison of weekly frequencies of cargo operations between China and the Netherlands from 2008 to 2010.

Source: www.caac.gov.cn.

in the near future. From the Netherlands' perspective, KLM, seeing the benefits of flying to Chengdu, was very keen to expand its network to more secondary Chinese airports and have, thus, chosen Hangzhou, the capital city of Zhejiang Province 100 kilometres south of Shanghai, for a prospective direct service.

With a view of providing as many commercial opportunities as possible for their businesses, the two authorities met in March 2010 and agreed to further remove the restrictions in terms of market entry and capacity permitting 68 weekly frequencies in total for each side split proportionally between passenger and cargo operations. In addition, both sides agreed to amend the designation clause, that is, the Chinese side accepted the community carrier, although with terms and conditions attached, which had been discussed between the two parties over the last couple of consultations without any success.

Table 5.7 provides a summary of the agreement.

Summary

Although China and the Netherlands only started their air transport relationship in 1996, it has developed smoothly and rapidly over the last 15 years. Restrictions on market entry, capacity and other business issues have been removed gradually, resulting in a more relaxed operational environment for carriers of both sides to explore the market opportunities (from one carrier for each side with two weekly frequencies to four carriers for each side with 68 weekly frequencies). Destinations of services have increased from one city in each country to numerous cities, with several intermediate and beyond points and full fifth freedom traffic rights.

TABLE 5.7 Summary of the agreement in 2010 for passenger and cargo operations

Items	Chinese side	The Netherlands side
Passenger routes and weekly capacity	33 in total, for points in China to the Netherlands	14 for Amsterdam–Beijing 14 for Amsterdam–Shanghai 15 for Amsterdam to other points including Chengdu and Hangzhou
Number of designated passenger carriers	2	2 (EU standard clause with respect to community carrier with terms and conditions)
Cargo routes and weekly frequencies	35 for cargo, but allows up to 49 including 14 out of the 33 passenger frequencies	35 in total for Amsterdam to points including Shanghai, Nanjing, Shenzhen, Guangzhou, and Hangzhou
Number of designated cargo carriers	2	2 (EU standard clause with respect to community carrier accepted with terms and conditions)

Source: Adapted from the 2010 ASA amendments.

Codeshare arrangements are permitted between both designated and non-designated carriers which have enabled the airlines to expand their services to offer a more coherent network connection. This progressive relaxation of restrictions on operations allowed the airlines to adapt their strategies and operations to meet the market demand, while at the same time enhancing the commercial and business links between the two countries.

China and the United Kingdom (UK)

UK and its air transport industry

As a contracting state to ICAO and the Party that concluded Bermuda I and II bilateral ASAs with the USA, which have significantly shaped the global air transport industry over the past six decades, the UK has played a pivotal role in that process. Although a small island off the European continent 35 kilometres away from France at the nearest point with a population of 62 million in 2009, the UK remains as one of the most important air transport hubs in Europe. The ACI connectivity report (ACI, 2018) ranked the UK the fourth most highly connected European country, behind France, Netherlands, and Germany (Figure 5.5).

Over the years, the passion for travel in search of international business opportunities and more sunny days has stimulated the rapid growth of the nation's air transport industry. In 1971, the British made around 6.7 million trips abroad compared to 66.4 million overseas trips in 2005. Of these, 43 million trips were made by air, compared with ten million by sea. By 2017, residents of the United Kingdom made 72 million overseas trips (Office for National Statistics 2018).

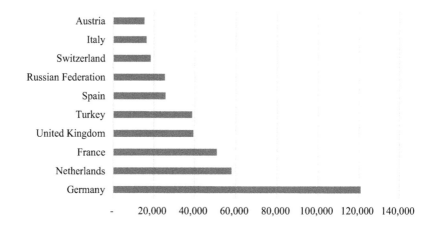

FIGURE 5.5 ACI Connectivity League, 2018.
Source: aci.aero.

The two major British carriers: British Airways and Virgin Atlantic Airways

The public enthusiasm for air transport has stimulated the British airlines to strive to remain competitive to grow and prosper. By the end of 2009, the UK hosted more than 30 airlines, with British Airways (BA) and Virgin Atlantic Airways (VG) (hereinafter referred to as Virgin) being world-renowned network carriers focusing on long-haul international routes while easyJet and Flybe developed innovative LCC models concentrating on intra-European markets. As a whole, the industry carried over 235 million passengers and over 2.3 million tonnes of freight in 2009 increasing to 285 million passengers and 2.6 million tonnes of freight by 2017 (DfT, 2017).

Although there can be little doubt that BA's history had been joyous from its inception, it has not been insulated from the types of challenges which have plagued other operators. Over time, the carrier has had to confront its own array of challenges, including financial woes, labour disputes, operational disruptions, and technological glitches. BA's premium product helped to establish it as one of the most profitable and favoured airlines the world over. Since the early 1990s when Lord King a radically re-shuffled the airline, it largely remained a very successful brand. Its most tumultuous year occurred in 2009 when, the airline reported its worst financial loss ever at a staggering £400 million—far greater than what analysts had expected. Occurring just after the 2008 financial crisis, the airline had to contend with rising fuel costs, the weakened British pound and declining market demand on a global scale. Since 2008, BA has experienced a radical overhaul of its corporate structure and operation, shedding thousands of jobs in an attempt to better its financial circumstances. Its financial woes were substantiated by labour disputes with unions, one of which was launched in 2010, where the airline was threatened with industrial action after proposing staff layoffs.

Despite the operational loss, and industrial action, the carrier remained one of the most competitive airlines in the world since the 1980s through periods of privatisation, mergers, and consolidation. In the early 2002s BA's second attempt to gain antitrust immunity to form an alliance with American Airlines was denied after the carrier refused to accept the precondition that it relinquish 16 slots at Heathrow. The application was also partially implicated by British regulators failing to reciprocate their American counterpart's enthusiasm for finalizing an open skies agreement. Antitrust immunity was eventually obtained in 2010 paving the way for an American Airlines, British Airways, Iberia joint venture allowing the carriers to cooperate, consolidating their US-EU market share. Focusing on high-yield business travellers, BA was able to remain as one of the top ten world airlines over the past few decades. It was ranked ninth in terms of both passengers carried and revenues made in 2009, recording 31.8 million passengers, realising 110.9 billion RPK and $12.8 billion revenues. However, this was less than half when compared with Lufthansa (after taking over Swiss International, Austria Airlines, and British Midlands Airways (BMI)), which registered revenues of $31.0 billion, followed by Air France-KLM with $29.7 billion, both of which are European-based network carriers competing with BA on a global scale.

The standalone situation was reversed in November 2010 when the proposed merger of BA with Iberia, Madrid-based Spanish legacy carrier, was approved by their shareholders. The merger paved the way for the development of one of the largest airline groups globally in the form of IAG (International Airlines Group) which serves as the parent company to European carriers Aer Lingus, British Airways, Iberia, Vueling, and its most recent addition, the LCC 'Level'). As the largest of the IAG group of airlines, BA continues to maintain a formidable position as a highly competitive global carrier. In the 2018 British Airways annual report, the carrier had a fleet of 294[3] serving more than 200 destinations with 45,000 employees. Data from the Official Airline Guide in 2018 revealed that the carrier's London Heathrow–New York JFK was the world's most lucrative air route by passenger revenue, amassing over US\$1 billion during the year April 2017–March 2018.

BA is one of the two UK-designated carriers operating between the UK and China. It inaugurated its twice weekly London–Beijing service in 1980 and expanded to Shanghai with five weekly frequencies in June 2005. According to OAG data, for the week commencing 29 April 2019 BA's scheduled operations were seven frequencies to Beijing competing with Air China's 21 and Hainan Airline's six. On the London-Shanghai route, BA operates ten frequencies against Virgin's seven and China Eastern's seven. In addition, BA's operations to Hong Kong with 14 weekly frequencies also faces competition from the UK-based competitor Virgin which offers seven frequencies, along with Cathay Pacific. Other routes served between China and the UK include Chengdu–Heathrow; Guangzhou–Heathrow; Shanghai–Manchester as well as other smaller cities with limited/ once weekly services.

Virgin was only launched in 1984 by Sir Richard Branson who established the Virgin brand through Virgin Records. Being committed to establishing a high-quality and value-for-money airline. Virgin has since grown rapidly offering scheduled services to 30 destinations around the world. With a 49 percent stake held by Singapore Airlines, Virgin operates through codesharing arrangements with numerous carriers including Aero Mexico Airlines, Air France–KLM, Air China, Singapore Airlines, Flybe, Virgin Australia, and Air New Zealand, and has remained independent of any global alliances. In 2013, Virgin Atlantic and Delta announced a partnership that would see Delta's total equity in the carrier rising to 49 percent. In spite of only having 46 aircraft in its fleet compared with BA's 294 by 2018, Virgin has been a strong competitor with BA on all long-haul routes where BA has an established presence. The airline maintained its competitive advantage by providing passengers with an unrivalled in-flight and airport experience such as limousine pick-ups and drive-through check-in, spa at the airport lounge, thus aiming to provide an unparalleled level of service in an ever increasingly stressful travel experience. Virgin Atlantic started its Shanghai service out of London Heathrow with two weekly frequencies in 1999 and has expanded to a daily service, competing with China Eastern and BA on the same route.

Airports in the UK

The UK has 56 airports with scheduled air services. The main municipal airports are usually separated by about 70 kilometres (Humphreys, 1999). Six airports are located within the London metropolitan area, namely Heathrow, Stanstead, Gatwick, Southend, Luton, and London City. Such a dense infrastructure provides the majority of the UK population with access to at least two airports within one and a half hours' drive time from their home, making air travel a very accessible transport mode for the British people. In addition to domestic traffic, the UK is particularly attractive to international passengers thanks to its convenient and extensive connecting network capabilities. Of the 292 million passengers handled by all UK airports in 2018, some 246 million were international passengers, representing 84 percent of the total (UK CAA, 2018).

The six London airports handled 60 percent of all passengers, with the rest going to regional airports. However, regional airports have enjoyed a faster growth in terms of passenger traffic between 1999 and 2015 (Figure 5.6), with terminal passenger numbers increasing in 2018 by 60 percent, compared with 43 percent at the six London airports.

Heathrow Airport is a well-established international schedule service airport for the UK, accounting for 27 percent of all passenger traffic in the UK followed by Gatwick with 16 percent (UK CAA, 2018). The airport's capacity is set to increase even further with the introduction of a third runway. Being the home base for BA and Virgin as well as the European hub for the Oneworld Alliance, Heathrow handled 80.1 million passengers in 2018, ranking seventh of the top ten global airports in terms of its passenger throughput. Of its total traffic, 94 percent (75.3 million) were international passengers, making it the second largest by this measure behind Dubai with 88.8 million (ACI, 2018). In 2018, Heathrow hosted 84 airlines connecting 203 cities in 84 countries with New York, Dubai, Dublin, Amsterdam, and Hong Kong being the top five most popular destinations.

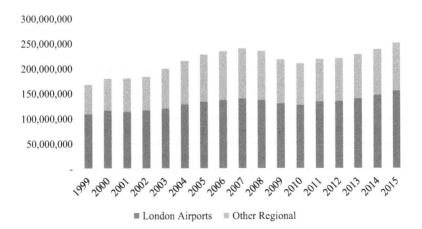

FIGURE 5.6 Number of terminal passengers at London and UK regional airports from 1999 to 2015.

Source: DfT.gov.uk.

Government agency responsible for international air transport and bilateral negotiations

The UK Department for Transport (DfT) is the government agency responsible for formulating international air transport policies. Reporting to the Department's Director General of Civil Aviation, the International Aviation, Safety and Environment Division (IASED) is responsible for the day-to-day management of bilateral aviation relations between the UK and its counterparts, which involves negotiating new bilateral arrangements and any amendments thereto, and enforcing existing bilateral arrangements. The division is also responsible for representing the UK's interests in EU-level negotiations.

In addition to DfT, the UK Civil Aviation Authority (CAA) is the UK's independent aviation regulator responsible for economic and safety regulations and consumer protection. The Economic Policy and Strategy team in the Economic Regulation Group oversee the development of policies in relation to a wide range of aviation matters. The Economic Policy and International Aviation team provides policy analysis and advice to the DfT on liberalisation and competition issues in the airline industry. This includes contributing to the development of EU aviation agreements, to the negotiations of bilateral ASAs, conducting economic studies and providing data and advice to UK competition authorities (specifically the Office of Fair Trading and the Competition Commission).

The team also plays a key role in cases involving the allocation of traffic rights between competing UK airlines. This kind of situation arises where more than one UK airline seeks to operate services between the UK and a country outside the European Common Aviation Area, whereby the bilateral air services agreement with that country restricts the number of services that can be operated, or the number of carriers that can fly. With the support of the CAA, the DfT has developed a mechanism which facilitates its formulation of negotiation strategies and tactics. This involves consultations with airlines, airports, and other stakeholders regularly, in particular before any bilateral talks, to ensure that the UK's interests are properly and effectively represented and protected in the negotiations (Coles, 2010; Humphreys, 2010; Knight, 2010).

The UK Government's view towards international air transport

Believing that the optimum allocation of resources of the common goods of society would be best served by a free market economy (Weber et al, 1947), the UK has led the way with its privatisation programme, which has long been suggested to influence economic policy throughout the world (Bishop et al, 1994). The full effects of privatisation can only be realised in a deregulated market structure where effective competition can flourish (Humphreys, 1999).

With respect to the air transport industry, the government believed that the interests of users will be best served if airlines are free to operate air services in competition with one another according to their commercial judgement, subject only to the application of normal competition policy. A liberalised market structure and reliance to the greatest extent possible on competition rather than regulation is perceived as the best way of delivering efficient aviation services and consumer benefits, which could only be achieved through the removal of all government-imposed restrictions on the commercial behaviour of airlines, putting aviation on the same footing as other industries. To this end, it is believed that the long-term interests of users would be best served by the full liberalisation of international aviation markets, involving the removal of all bilateral restrictions so that the airline industry can compete equitably (UK CAA, 2008).

Such a liberalised view towards air transport has enabled the UK government to actively pursue liberalisation both domestically and internationally. The notion of transforming its domestic airline industry was initiated in 1967 with the release of the Edwards report, which argued that the airline industry should be regarded as a commercial enterprise. Its purpose should be to see that each customer gets what he wants, not what somebody else thinks he might want, at the minimum economic price that can be contrived (Staniland, 2003). The regulatory change affecting the airlines occurred in 1979 under the Thatcher Administration, which was determined to transform Britain from a mixed economy to a market economy. Although the process of consolidating the national carrier, BA, had not been completed until 1987, it was still well ahead of the restructuring and privatisation of Air France and Lufthansa which only took place in the mid-1990s (Staniland, 2003). By passing the Airports Act in 1986, the UK also privatised its airport sector and transformed all of its municipal airports into commercial companies with an objective of encouraging enterprise and efficiency in the operation of major airports to allow the introduction of private capital rather than being subsidised by taxpayers or ratepayers.

Over the years, the UK government has been persistently pursuing liberalised bilateral arrangements with its counterparts to eliminate all unnecessary restrictions to create a pro-competitive environment for the UK airlines to operate and to increase the benefits for consumers. By November 2010, the UK had concluded more than 150 bilateral ASAs. Among them, four are regarded as

> truly liberalised arrangements with unlimited 3rd, 4th, 5th, 6th, 8th and 9th traffic rights, (including cabotage), without any restrictions on the points that can be served in either country. In some cases the traditional ownership and control/nationality requirements have also been liberalised.
>
> *(Knight, 2010)*

Another 27 are considered

> open ASAs/bilateral relationships, which means there are no restrictions on the 3rd and 4th traffic rights between the two countries, nor any restrictions

on the points that can be served in either country. Although the majority of these arrangements have limited 5th and 6th traffic rights arrangements, quite a few do have unlimited 5th and 6th traffic rights such as those with the US and Canada.

(Knight, 2010)

In 2018, the government of the UK through the DfT made available for consultation a strategic plan for the development of the UK's aviation industry towards 2050. Proclaiming the UK to be home to the third largest aviation network globally, it declared that further expansion and liberalisation of the industry is critical to enhancing Britain's global connectivity. To achieve this end, collaborative efforts with ICAO to propagate the benefits of liberalisation along with renewed efforts to ensure commercial flexibility and limit restrictions imposed through its ASA were deemed strategic aims. The policy advocates that the UK's overarching objectives towards enhancements in global connectivity are to (a) maintain and improve the UK's connectivity, (b) seek more liberalised arrangements, and (c) improve transparency of aviation agreements. To build what the government envisions as a globally connected Britain, these goals are driven by various strategies including enhancing trade, supporting robust aviation security frameworks, and strong environmental standards among others. Table 5.8 provides an extract of some of the stipulated strategies.

TABLE 5.8 Areas of focus for a globally connected Britain—Aviation 2050: the future of UK aviation

Focus	Strategies
Environment	• Environmental emissions reduction measures; • Support Stronger Carbon offsetting and Reduction Scheme for International Aviation (CORSIA); • Negotiate longer-term international aviation climate emissions goals;
Security	• Support ICAO Global Aviation Security Plan; • Increase support for nations through Universal Security Audit Programme;
Trade	• Promote benefits of and encourage liberalisation; • Reduce burdens in ASAs and enhance commercial flexibility
Technology and Innovation	• Support international development of more responsive global regulations; • Encourage accountability and transparency in international standards; • Promote engagement between ICAO and industry to assess regulatory impacts;
Bilateral leadership	Provide training and support to other bilateral partners to enhance their capacity
New EU–UK relationship	To establish the most liberally practical arrangement with EU

Focus	Strategies
Air traffic rights	• Fully liberalise traffic rights and remove freedoms, frequencies, destination, and carrier restrictions; • Develop multilateral agreements;
Airline ownership and control	Replace ownership restrictions and pivot towards primary place of business as measures of control
Interchange	• Facilitate interchange agreements through an apt regulatory environment; • Encourage the development of adequate policies in consult with the CAA; • Promote liberalisation at bilateral and multilateral levels; • Review ASAs after 10 year with outlooks towards further liberalisation as necessary;
Transparency	Publish relevant extracts of ASAs online (traffic rights, frequencies, etc.)
Government's bilateral strategy	Extend the UK's share in global market for exports of air transport services and aerospace goods
Barriers to trade	In consultation with the British Aviation Group, to establish a new Aviation exports board with greater intersectoral collaboration
Free trade agreements	In FTA negotiations reduce barriers for the export of auxiliary UK aviation services (e.g. computer reservation system services, airport operation goods and services)

Source: DfT.gov.uk.

Bilateral air transport relations between China and the UK

The first China–UK bilateral ASA was signed in November 1979 with immediate effect. Ever since then, more than ten rounds of discussions have been conducted with the latest being held in February 2016. This agreement precipitated a drastic increase in potential capacity between the two nations with weekly frequencies being raised from 40 to 100 each way and limits on the Chinese cities eligible to be served by UK airlines removed.

Arrangements before 1997

Like other bilateral ASAs negotiated in the 1970s, the China–UK agreement was in essence very restrictive and protectionist. The air link between the two countries carried more of a political implication than an economic and trade significance. The arrangements allowed both sides to designate one carrier to operate two weekly frequencies between the capital cities, namely Beijing and London, respectively, with a couple of intermediate points enroute including in South Asia, the Middle East, and Europe due to the technical capabilities of aircraft during this period (Table 5.9).

TABLE 5.9 Key points in the 1979 Agreement

Items	Chinese side	The UK side
Number of designated carriers	1	1
Weekly capacity	2	2
Passenger and cargo routes	Beijing–intermediate points–Frankfurt–London	London–intermediate points–Beijing

Source: Adapted from the 1979 ASA.

The carriers of both sides, for example, BA and the CAAC (changed to Air China in 1985), soon launched their two weekly services, respectively, in November 1980, though BA flew out of its Heathrow home base while the CAAC landed at Gatwick (moved to Heathrow in the late 1980s). Due to the small volume of bilateral trade and few people-to-people exchanges, the market demand had remained weak over the following two decades, resulting in little or no desire of both parties to expand the services. Although negotiations were held, respectively, in 1985, 1990, 1991, and 1994, they mainly focused on the revision of the performance of the carriers rather than the removal of restrictions. The consultations held in 1991 and 1994 were primarily devoted to the arrangements between mainland China and Hong Kong and the beyond points for the expansion of air links.

The situation remained unchanged until June 1996, one year before the handover of sovereignty of Hong Kong to the Chinese government. The agreement reached then allowed both parties to increase to a third and fourth weekly frequency from 1996 and 1997, respectively (Table 5.10), and was the first relaxation of the restrictions on operations since 1979, almost 18 years since the ASA became effective. The 1996 arrangement also amended the route schedules allowing the Chinese carriers to overstop at Hong Kong enroute to London with full traffic rights.

TABLE 5.10 Agreement between China and the UK in 1996

Items	Chinese side	The UK side
Number of designated carriers	1	1
Route	Beijing–Hong Kong–London	London–Beijing
Weekly capacity	3 as of winter 1996	3 as of winter 1996
	4 as of summer 1997	4 as of summer 1997

Remarks: The Chinese side was allowed to operate Beijing–Hong Kong–London with a maximum of four weekly frequencies with full traffic rights. However, an overstop at Hong Kong was only allowed one way. Capacity was included in the total entitlement.

Source: Adapted from the China-UK 1996 ASA amendments.

Arrangements after 1997

The year 1997 carries great significance in the bilateral relations between China and the UK. In this year, the sovereignty of Hong Kong was passed back to the Chinese government, which meant that the central government in Beijing was able to exercise its jurisdiction over Hong Kong on its dealings with the international community, although the Hong Kong SAR has the authority to conclude any bilateral treaties with the approval of the central government.

As of 1 July 1997, all arrangements agreed between China and the UK with respect to mainland China–Hong Kong and beyond points, except Beijing–Hong Kong–London arrangements, were no longer an integral part of the Sino-UK package, but rather the special arrangements between the Chinese government and the Hong Kong SAR authority (Hong Kong Basic Law, Articles 132–133,[4] 1990). The separation of the element of Hong Kong enabled both governments to concentrate on the country-pair markets, which was reflected in the discussions held respectively, in 1998 and 2004.

Discussion and arrangement reached in 1998

The arrangements reached in 1996 enabled carriers of both sides to increase their weekly frequencies. Both BA and Air China started their third weekly services from the winter season of 1996. By 1998, BA had operated four weekly frequencies on the London–Beijing route while Air China remained at three weekly services.

The financial crisis in Southeast Asia in 1997 had a significant impact on Chinese carriers, resulting in a weak demand for international air transport in the Southeast Asia market. However, it did not discourage Sino-UK officials from scheduling a discussion in October 1998, the time when the then British Prime Minister Tony Blair paid a visit to China. Both parties agreed that frequencies on the Beijing–London route should be increased from the current four weekly to five as of summer 1999 and six as of summer 2000. The British side was very keen to have Shanghai as a second destination in mainland China and would want a second carrier to be eligible to operate in the market. The two thus agreed a deal allowing two carriers from each side to operate in the market and Shanghai being included as a second point of call for the UK carriers with two weekly frequencies as of summer 1998 and up to four weekly frequencies as of summer 2000.

The new traffic rights available, unfortunately, could not satisfy the appetite of the UK carriers, with both BA and Virgin wishing to exercise the country's full entitlements of the two weekly services to Shanghai. To decide whether to award the two airlines one weekly frequency each or award both frequencies to one of the airlines (CAA, 1998), the UK CAA convened an open hearing with relevant stakeholders being present.

Believing that Shanghai was one of those few cities with exceptional commercial potential in the coming decades for international business in general and air travel in particular (CAA, 1998), both BA and Virgin presented a strong case with an attempt to convince the CAA that each was more appropriate than the other for the new operation. BA, already operating to Beijing four times per week, stated that the decision as to which airline should fly to Shanghai should be made primarily on the basis of the reasonable interests of users, with further weight given to the need to facilitate competition by British airlines with other airlines rather than competition between British airlines, securing the sound development of the industry and to the economic prospects of the services. Virgin, on the other hand, argued that the award to Virgin would be consistent with the authority's policy to ensure active competition between British airlines while at the same time benefiting the travellers by giving them a second choice to fly into mainland China (CAA, 1998).

After considering the arguments from both sides as well as user group representatives, the CAA eventually decided to award the two weekly Shanghai services to Virgin rather than BA, giving the rationale that competition in the UK industry and in the UK/China market would be strengthened to a greater degree with a new entrant into the market (CAA, 1998).

Virgin, thus, became the second carrier in the UK–China market in February 1999 and inaugurated its London–Shanghai services in May. For Virgin, the Shanghai route was symbolic and significant, which meant that it was able to compete with capacity on international long-haul routes where BA had not been established (Humphreys, 2010). BA, on the other hand, did not give up its ambition to expand its services into Shanghai. It lodged another application to CAA in 1999 for the other two weekly frequencies that became available from summer 2000. However, after another open hearing in November 1999 convened by CAA, these routes were also awarded to Virgin giving the carrier all four weekly frequencies to Shanghai (Humphreys, 2010).

For the Chinese side, the allocation of the traffic rights with respect to Shanghai was less contentious as only China Eastern expressed a wish to become the second carrier in the China–UK market. The airline, though receiving the approval for a Shanghai–London service shortly after the negotiations, failed to launch the flight until 2003, almost four years later than its British counterpart.

Consultation and outcome agreed in 2004

Following the discussions in 1998, bilateral trade between China and the UK picked up rapidly. The exchange of visits of senior officials further inspired economic and trade activities between the two countries. Following the successful visit of Tony Blair to China in October 1998, Chinese President Jiang Zemin paid his first visit to the UK in 1999 and launched a China–UK Forum, with an aim of enhancing the bilateral relationship to a comprehensive partnership. In 2001, the trade value between the two countries had surged to $10.3 billion, a 4.1 percent increase over that in 2000. Before 2003, the UK had remained China's second largest trade partner of the EU member states though later replaced by the Netherlands.

Passenger traffic increased rapidly after 1997. According to CNTA, the UK recorded 302,500 arrivals in China in 2001, overtaking Germany and France. This compared with 227,900 in 1997, representing an average 7 percent increase year on year. On the other hand, Chinese visitors to the UK had increased remarkably recording 88,000 in 2001, a 40.2 percent increase over the previous year. In particular, the UK became the second largest destination for Chinese students in the early 2000s just after the USA, with tens of thousands flocking to the country each year. Figures 5.7 and 5.8 summarise the air passenger traffic between China and the UK between 2001 and 2003, compared with other major western countries in Europe, which reveal that Chinese outbound visits to the UK enjoyed the most rapid growth during the period compared with the UK arrivals in China.

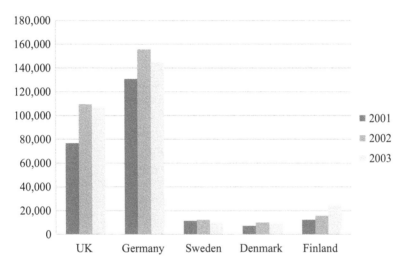

FIGURE 5.7 Chinese outbound visits to 5 EU member states between 2001 and 2003. *Source*: CNTA.

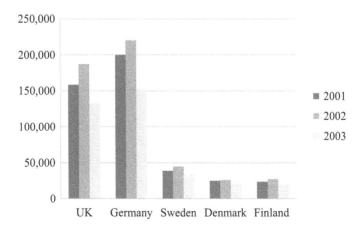

FIGURE 5.8 Arrivals from 5 EU member states to China between 2001 and 2003. *Source*: www.cnta.gov.cn.

The booming of bilateral trade and passenger traffic encouraged both parties to increase their services. In the winter season of 2003, Air China was operating five weekly services between Beijing and London compared with BA's three weekly services on the same route. Virgin flew four weekly services between London and Shanghai, while China Eastern inaugurated three weekly services in the summer season of 2003. Compared with the entitlements agreed in the 1998 arrangements, five out of six of the Beijing–London route frequencies were used by Air China, while BA only used half of its traffic rights. For the Shanghai–London route, Virgin had used up all of its four weekly entitlements while China Eastern only used up 75 percent of its entitlement. The effective use of capacity was also reflected in the number of passengers carried, which had seen a steady growth over the years (Figure 5.9). Apparently, since 2001, Air China had both provided more flights and carried more passengers than BA to and from Beijing, although BA carried more business passengers which made up 65 percent of its total traffic compared with 19 percent for Air China[5] (CAAC, 2007).

The high demand and satisfactory performance inspired both parties to seek more traffic rights. Air China saw significant potential in the Sino-UK market and requested a daily service on the Beijing–London route as well as an entry into the Shanghai market, quoting the magnificent increase of passenger numbers boarding its aircraft with more than 70 percent load factors on average between 2001 and 2004 (except a certain period in 2003 due to the effect of SARS). China Eastern, though recently joining the competition with Virgin on the London–Shanghai route, was also very optimistic about the continuous high demand in the market and wanted to enter the Beijing–London market, although

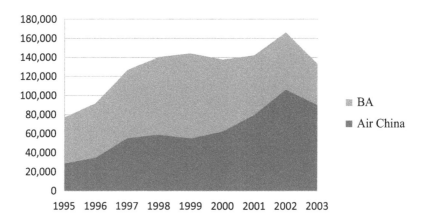

FIGURE 5.9 Traffic carried by Air China and BA between 1995 and 2003.
Source: CAA (2007).

it was quite conservative about cargo operations. In addition to the requests from incumbents, the market attracted more Chinese carriers, such as China Southern, Hainan Airlines, Shanghai Airlines, and China Cargo, who were all keen to become designated carriers on the Sino-UK market.

The requests of the Chinese side were echoed by their British counterparts. The UK proposed to separate the operations of passenger and cargo services and remove the restrictions on the numbers of designated carriers with capacities being increased. In addition, the British delegation wished to amend the designation clause in the existing ASA to bring it into conformity with the EU law. With identical requests and confidence in the prospective market growth, the 2004 consultation went smoothly with an agreement reached to both parties' satisfaction (Table 5.11).

Compared with the agreement of 1998, the 2004 arrangements were a major step in terms of relaxing the restrictions on the operations. The number of designated carriers was completely relaxed from 2 for each side to unlimited. The capacity was increased from 10 weekly for each side (6 for Beijing and 4 for Shanghai) to 20 weekly in 2004 and 38 in 2006, which was almost a four fold increase. In 1998, there was no specification on cargo operations, which were treated as part of passenger services, while in 2004, cargo operations became independent from passenger operations with no restrictions on capacity.

The agreement enabled BA to increase its operations to five frequencies to Beijing while at the same time to enter into the Shanghai market, which it started in June 2005 that it had longed for since 1998. Virgin was able to invest more in its well-established Shanghai market to increase its services to a daily operation. For the Chinese side, Air China was allowed to increase its frequencies to London with its cargo arm starting all-freight services to Manchester. China Southern was awarded traffic rights to the UK, although it has had no specific date to launch the service yet. Table 5.12 shows the traffic between London Heathrow and Beijing in the first eight months of 2004.[6]

TABLE 5.11 Arrangements agreed between the two parties in 2004

Items	Agreement
Routes for passenger and cargo	Separation of passenger and cargo operations
Number of designated carriers	No restrictions for both side
Capacity for both sides	2004: 20 weekly frequencies
	2005: 31 weekly frequencies
	2006: 38 weekly frequencies
Fifth freedom traffic rights	Selected intermediate and beyond points with full fifth freedom traffic rights

Source: Adapted from the 2004 ASA amendments.

TABLE 5.12 Heathrow–Beijing scheduled services for the first eight months in 2004

	Point to point passenger	Flights	Total passengers	Seats	Seat factor (%)
Air China	108,979	561	108,979	170,406	64
BA	73,697	384	73,697	146,136	50
Total	182,676	945	182,676	316,542	58

Source: CAA (2007).

Discussion and arrangement reached in 2011

Until 2007, international air transport had seen magnificent increases in demand, with a historical record of revenue passengers carried, before being followed by an unprecedented economic downturn which hit the global airlines severely in 2008 and 2009. Before the industry was able to recover it was again impacted by, the swine flu epidemic across the Americas and the volcanic ash in Europe, with which caused to plummet on a global scale. LCCs also started to spread to Asia including China with air travel becoming an affordable transport mode for the public.

In terms of the general regulatory environment, the EU and the USA had concluded their second stage open skies agreement to create an OAA, enabling all EU established carriers to operate services from any points in the EU to any points in the USA. At the bilateral level, the UK had concluded various few open skies agreements with other countries, including India, Singapore, Canada, Australia, and New Zealand, which consequently stimulated a significant growth in its international air transport market. A UK CAA study conducted in July 2007 on the UK's long-haul air transport market (excluding the EU market) revealed that the open skies agreement covered about 60 percent of traffic (UK CAA, 2007), which has significantly stimulated the traffic growth. British airlines have taken advantage of such a relaxed regulatory environment, with BA having launched a subsidiary open skies in Paris for direct cross-Atlantic services. BA itself has merged with Iberia to form another mega-carrier in Europe and its alliance arrangements with American Airlines have received anti-trust immunity from the US DoT.

From China's perspective, the Chinese government gained more experience in negotiating liberalised agreements with its counterparts, especially with a handful of EU member states, and have accepted the establishment and nationality clause and removed constraints on cargo operations and fifth freedom traffic rights for passenger services. Chinese airlines have grown rapidly with more traffic carried and revenues improved. Air China has become a Star Alliance member while both China Eastern and Southern are with SkyTeam.

Bilateral trade has grown remarkably, with more passengers travelling between the two countries. Figures from CNTA showed that British inbound travellers to China increased from 500,000 in 2005 to 551,500 in 2008, although dropping to 528,800 in 2009 due to the economic downturn. However, China was still ranked 39th among long-haul country destinations from London Heathrow Airport in 2008, though Hong Kong SAR ranked 19th, recording 1.55 million English arrivals from London Heathrow (RDC aviation, 2010).

Looking to the forthcoming consultation, the UK side was very confident that more relaxed arrangements would be agreed so as to promote its economic growth at regional airports (Knight, 2010). They also hoped to resolve the slot issue in major Chinese airports so as to allow UK carriers to serve the market more efficiently. The UK was also interested in some arrangements with respect to the traffic between Hong Kong and Taiwan so as to allow its carriers to grow in this market. They hoped that the designation clause would be agreed by the Chinese side so that it will fulfil its obligations as an EU member state. The Chinese side hoped that slots could be available at Heathrow Airport to allow Chinese carriers to compete and was ready to discuss the designation clause. Until 2012, UK carriers had consistently led the market in the number of weekly frequencies to China. By 2014, the Chinese carriers had only utilised approximately half of their available capacity which when coupled with the growing outbound Chinese tourism demand positioned the market for further growth. The UK and China revised the Air Service Agreement increasing the approved flight levels from 40 to 100 weekly flights in an amendment in 2016.

What the future holds

On 23 June 2016, citizens of the UK voted in a referendum to renounce the UK's membership with the European Union which is likely to have implications for air services between the UK and China. This referendum commenced a protracted period of negotiations intended to give effect to the UK's orderly exit and the unbundling of decades of EU-guided laws, processes, and practices. Already one of the most geopolitically charged industries, the UK's membership within the EU had overtime consolidated a significant amount of EU influence on the UK aviation policy—as with any other EU member state. During its tenure within the EU, the EU commenced horizontal negotiations with 41 countries which the UK would have been subject to given its membership in the EU. Of these horizontal negotiations, in 2015, the EU and China began discussing areas of cooperation in aviation, after a letter of intent was signed to do so in 2013 and the agreement was subsequently initialled in 2017 as previously examined. The horizontal agreement which signalled China's recognition of the principle of EU principle designation thus created prospects for airlines registered in an EU member state to capitalise on market access otherwise negotiated between China and other EU member states. There is little doubt that the horizontal agreement between China and the EU will extend traffic and trade opportunities between the bloc and China. From the UK's perspective however, given its impending departure from the EU, it still stands to be seen where the China–UK aviation relationship lies after its departure from the bloc in light of China's ongoing positioning towards foster a more resilient China–EU partnership.

Notes

1 Great Wall temporarily suspended its five weekly Shanghai-Beijing-Amsterdam-Seoul service in winter/spring season 2006.
2 The Netherlands investment in China was $617 million in 2007 and $862 million in 2008.
3 BA had 233 aircraft in the fleet in 2009.
4 **Article 132:** All air service agreements providing air services between other parts of the People's Republic of China and other states and regions with stops at the Hong Kong Special Administrative Region and air services between the Hong Kong Special Administrative Region and other states and regions with stops at other parts of the People's Republic of China shall be concluded by the Central People's Government.

   Article 133: Acting under specific authorisations from the Central People's Government, the Government of the Hong Kong Special Administrative Region may:
   1 renew or amend air service agreements and arrangements previously in force;
   2 negotiate and conclude new air service agreements providing routes for airlines incorporated in the Hong Kong Special Administrative Region and having their principal place of business in Hong Kong and providing rights for over-flights and technical stops; and
   3 negotiate and conclude provisional arrangements with foreign states or regions with which no air service agreements have been concluded.

5 CAA airport statistics reduced by 6 percent to take account of non-revenue passengers (CAA, 2007).
6 CAA airport statistics reduced by 6 percent to take account of non-revenue passengers (CAA, 2007).

References

ACI., (2018), Airport Industry Connectivity Report 2018, available at, https://www.aci-europe.org/downloads/resources/ACI%20EUROPE%20Airport%20Industry%20%20Connectivity%20Report%202018.pdf

Bishop, M., Kay, J., Mayer, C., 1994. Privatisation in performance. In: Bishop, M., Kay, J., Mayer,.C. (Eds.), Privatisation and Economic Performance, pp.1–14, Oxford University Press, Oxford

CAA, (1998), Decision on air transport licences and route licences, decision of the Authority on its proposal to vary licence 1B/10 held by British Airways Plc and licence 1B/35 by Virgin Atlantic Limited heard on 10 December 1998, 5/98, available at www.caa.co.uk/publicationsCAAC, (2005), *Statistical data on Civil Aviation of China*, Civil Aviation Press of China, Beijing.

CAAC, (2006), *Statistical data on Civil Aviation of China*, Civil Aviation Press of China, Beijing.

CAAC, (2007), *Statistical data on Civil Aviation of China*, Civil Aviation Press of China, Beijing.

Coles C., (2010), interviewed conducted in December 2010 via telephone.

Daniels A., (2010), interview conducted via email communication.

DfT, (2017), UK aviation forecasts, 2017, available at https://assets.publishing.service.gov.uk/government/uploads/system/uploads/attachment_data/file/781281/uk-aviation-forecasts-2017.pdf

Eurostat News Release, (2004), EU-China Summit, China now second trade partner of EU25, Trade doubled between 1999 and 2003, (7 December 2004), available at http://europa.eu.int/comm/eurostat/

Hong Kong Basic Law, (1990), available at www.basiclaw.gov.hk/en/basiclawtext/chapter_5.html

Humphreys B., (2010), Telephone interview conducted on 8th December 2010 in the UK.

Humphreys I., (1999), Privatisation and commercialization, changes in UK airport ownership patterns. *Journal of Transport Geography*, 7, 121–134.

Knight S., (2010), Telephone interview conducted on 02 December 2010 in the UK.

Liang N., (2010), Interview conducted in Nov 2010 in Beijing.

Ministry of Infrastructure and Environment, the Netherlands, (2009), White Paper on Dutch aviation: competitive and sustainable aviation sector for a robust economy, April, 2009, available at www.verkeerenwaterstaat.nl/english

Office for National Statistics, (2018), Travel trends: 2017, available at www.ons.gov.uk/peoplepopulationandcommunity/leisureandtourism/articles/traveltrends/2017

Official Airline Guide, (2018), Billion dollar route, available at www.oag.com/blog/billion-dollar-route

RDC Aviation, (2010), UK passenger & cargo monitor, Analysis of UK passenger & cargo flow, Jan–Dec 2009, available at www.routepro.net

Rowland van K., (2010), Interview conducted at Amsterdam Airport in July 2010.

Schiphol Group, (2009), Schiphol Airport Annual Report 2009, available at www.annualreportschiphol.com/

Schiphol Group, (2019), Schiphol Airport Annual Report 2019, available at: https://www.annualreportschiphol.com/Shanghai Business Review, (2009), Flying high: interview with Mr. Mark Arxhoek, regional sales director of Air France-KLM Greater China, Volume 6: Issue 9, October 2009, available at www.sbr.net.cn

Staniland M., (2003), Government birds, air transport and the state in Western Europe, Rowman & Littlefield Publishers, Inc., Lanham, MD.

Sweijen W., and Pouwels B., (2010), Interview conducted in Amsterdam Airport in July 2010.

UK Civil Aviation Authority, (2008), CAA statement of policies on route and air transport licensing, available at www.caa.co.uk/docs/589/ERG_EPIA_Statement_of_Policies_Aug_2008.pdf

UK Civil Aviation Authority, Economic Regulation Group, (2007), Connecting the continents, long haul passenger operations from the UK, CAP771, (July 2007), available at www.caa.co.uk/publications

UK Department for Transport, Managing Bilateral Arrangements, (n,d), available at www.dft.gov.uk/pgr/aviation/international1/managingbilateralarrangements

UK CAA., (2018), Airport Data, available at https://www.caa.co.uk/Data-and-analysis/UK-aviation-market/Airports/Datasets/UK-Airport-data/Airport-data-2018-01/

Weber M., Henderson, A.M., and Parsons, T. (1947),, *The theory of social and economic organizations* (1st Amer. edition) Oxford University Press.

Williams A., (2009), The externalisation of air transport reform in Europe: a selective analysis of the developing role of the European Commission, *Aviation Education and Research Conference Proceedings*, Blenheim, New Zealand, July 2009, pp. 44–55.

Wong S., (2009), Tulips bloom in China, Trade between the Netherlands and China is growing steadily, with a special emphasis this year on the food, water and health sectors, *Shanghai Business Review*, October 2009, Volume 6: Issue 9, available at www. sbr.net.cn

6

CHINA, JAPAN, AND SOUTH KOREA

Overview: economic relations between China, Japan, and South Korea

Thanks to the reforms of China's trade policy since the late 1970s, the economy has rapidly developed to become the world's second largest and the world's largest with respect to purchasing power parity (Purdie, 2019). Similarly, Japan's stable economic growth in recent years has undercut analysts' expectations of contractions for the world's third largest economy, while economic expansion in South Korea has been fuelled by sustained development. Although the three nations each have implemented free trade agreements (FTAs) with major trading blocs such as the USA and the EU, they are yet to finalise prolonged negotiations on a China–Japan–South Korea FTA between themselves. The three nations embarked on the journey towards economic harmonisation with a joint study issued in 2011 and the first round of negotiations commenced in 2013.

Notwithstanding Japan, China, and South Korea's GDP accounting for almost 20 percent of the global GDP, 18 percent of the world exports and 16 percent of imports, prospects still exist to enhance economic integrations between the three nations (MOFA, 2011). As evidence of the potential economic integration which could be achieved, while intra-regional trade between the trio increased from 12.3 percent in 1990 to 21.7 percent by 2008, this was still far lower than the intra-regional trade levels in regions such as the EU which stood at 60 percent. The complimentary economic composition of the three nations has been a driving force of arguments supporting the establishment of the free trade area between the three countries and China's accession to the WTO had made it more amendable to such agreements.

By April 2019, the trio had entered their 15th round of negotiations on the establishment of the FTA. In the six years since the commencement of the trilateral negotiations, the three nations still face deep-seated issues to be resolved which stem from foreign policy positions in the case of territorial disputes, previous wars, and similar pillars of economic growth. While trilateral negotiations remain unresolved, bilateral discussions amongst themselves have been more progressive. The remainder of the chapter addresses these bilateral China–Japan and China–South Korea relations respectively and discusses their impacts on the air transport environment between each of the two nations.

China–Japan relations

China and Japan normalised their diplomatic ties in 1972. Both being powerful and affluent in Asia, the two are undergoing domestic transformations that have had major regional and global economic consequences. As of 2003, the two economies already constituted 47.9 percent of the value of merchandise trade in Asia and 12.5 percent globally (Pekkane and Tsai, 2005). Bilateral trade rose from $18.2 billion in 1990 to $236.6 billion in 2007 or 17.7 percent of total Japanese external trade, with Sino-Japanese trade exceeding that of the American-Japanese, making China the largest trade partner of Japan ever since then (Xing, 2008). As recently as 2017, Sino-Japanese trade had grown even further to $297 billion (representing 19 percent) of Japan's exports.

The economic relationship between China and Japan presents one of the most significant geopolitical relationships in Asia and on a broader global scale. Since the first Sino-Japanese 1890s war, the two have continued to seek to dominate geopolitical leadership in the region (Stuart, 2014). While trade between the two nations represents the third largest bilateral trade partnership globally, China and Japan face tenuous historical and political relations. Notwithstanding, the structure of the two production-led economies has created a symbiosis which has allowed China to benefit from Japan's competence in industrial technology whereas Japan was able to use China as a source of economic growth in the 1990s to propel its economic recovery through an external-led growth and development strategy (Armstrong, 2012). Japan's economic resurgence was so strongly supported by the growth in the Chinese economy that the International Monetary Fund (IMF) attributed almost all of the increase in Japanese trade in Asia during this period to exports to China (Chung, 2014). The growth in China and its consequential growth in consumer demand thus spurred Japan's economy. In the period between 2000 and 2004, Japanese exports grew by 8.5 trillion yen, 4.7 trillion of which were exports to China, equating to 55 percent of Japanese exports. While Japan's largest trading partner has been China, Japan only represented China's fourth largest trading partner. According to the World Bank, in 2017 Japan's total trade deficit with China totalled $28 billion US dollars with Japanese exports to China summing $137 billion compared with $165 billion in imports (Figure 6.1).

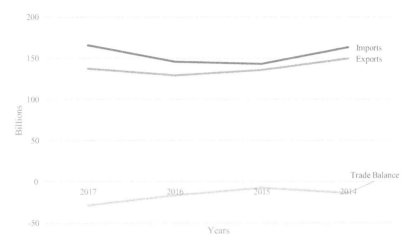

FIGURE 6.1   China–Japan trade balance.
Source: worldbank.org.

Major exports from Japan to China include electric–mechanical products, transportation goods, metals, and chemicals. In 2017, machinery and electrical products accounted for $54 billion of the $132 billion in Japanese exports to China. Similarly, exports by China to Japan also principally included electric–mechanical products, transportation goods, metals, and chemicals. Of the $165 billion in exports to Japan, $75 billion related to electric–mechanical products. Trade in goods between the two nations is reflective of the expanding economies. In 1992, clothing and textiles were the largest exports from China to Japan accounting for almost a third of the trade value (Chan and Kuo, 2005). As a consequence of Chinese economic growth, Japanese investors exploited business opportunities and took advantage of lower labour costs in China to produce goods often exporting them back to Japan. The prevalence of Japanese investment in China therefore helped to extend China's trade surplus with Japan.

As in any bilateral trading system, however, economic trade between nations is also closely tied to political relations. China's overarching perspective has been that Japan has been heightening its foreign policy stance in relation to disputes between the two, particularly over territorial claims. However, as Stuart (2014) has asserted, China relies on enforcing its governmental sense of legitimacy and stability to sustain its economic prosperity and propel its global dominance and overtake the USA as the world's superpower. As Japan is an integral trading partner, the pursuit of this goal stands to be undermined by continued tensions between these two nations.

In the modern global economy, China however holds the economic power relative to its neighbouring Japan. This is in contradiction to the earlier trade dynamics in the region where Japan represented the fastest growing economy in the region in the 1950s and 1960s. However, the economic reforms in China provided sustained economic growth, and since Japan's economic recession in the 1980s, China has maintained its advantage.

Japan's international air transport industry

Overview

With a population of 126 million and a landmass of 365,000 km^2, Japan plays a strategic role in the ASEAN air transport policy arena. Soon after World War II, civil aviation in Japan was restricted with Japanese airlines only being permitted to establish services in the 1950s. Consequently, in 1951 Japan Airlines was founded as a private carrier before being subsequently subsumed as a semi-governmental organisation in 1953 in attempts to expand its international competitiveness. In 1958, two previously formed domestic carriers were merged by the Japanese government to prevent their collapse, ultimately leading to the creation of ANA (All Nippon Airways). Japan Airlines and ANA jointly serve as two of Japan's oldest carriers.

International air transport in Japan has increased rapidly since the opening of Narita International Airport in 1978 which focuses primarily on international traffic. However, during the evolution of the industry, there have been various interventions which have affected the stability of international passenger arrivals. In September 1994, another major gateway was established as Kansai International Airport with approximately 12 million passengers.

Until the September 2001 terrorist attacks on the USA, Japan witnessed annual international air passenger arrivals of 52 million. Through 2003, the Iraq War/SARS pandemics led to further declines with total international passengers falling to around 45 million after which they then recovered to 55 million, setting a new high. By the opening of Central Japan International Airport in 2005, a further increase to approximately 57 million international passengers was witnessed. However, this resurgence was only to be short lived after the global financial crisis in 2008, which reversed the steady uptrend, resulting in international passengers decreasing to around 53 million. By 2012, international passenger arrivals to Japan had only recovered to reach a new peak at 60 million.

The periods of 2000–2004 and 2008–2009 represent some of the most unstable growth periods of the 21st century for international air transport around the world. However, while the industry was beginning to stabilise by 2011, international air transport growth in Japan was again stunted when tragedy struck in the form of the Great East Japan Earthquake and tsunami. The natural disaster was estimated to have claimed 20,000 lives, and resulted in $210 billion in damage (World Bank, 2012). Coastal Sendai airport was most significantly impacted by the disaster requiring reconstruction efforts with damages including a collapsed roof. The interconnected nature of transport system, however, had a ripple effect which also impacted the resurgence in international air transport. Up to

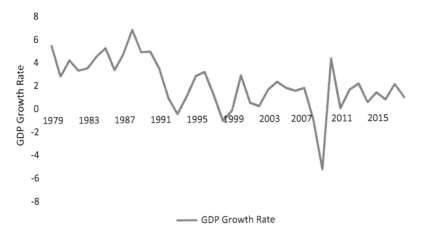

FIGURE 6.2 Japanese GDP growth rate, July 2019.
Source: worldbank.org.

one month after the earthquake, 22 railways including the Shinkansen remained inoperative, and numerous other land transport arteries were affected, including 171 national roads, 20 highway routes, and over 536 local roads and various bridges (Leelawat, 2015). Additionally, the ensuing tsunami from the earthquake flooded and affected the Fukushima nuclear power plant causing a meltdown of the nuclear reactor. Consequently, international interests in travelling to Japan immediately waned with governments such as the US DoS issuing travel advisories. In 2011, Japan's GDP growth fell to 0 percent and ranged between 1.5 and 2 percent between then and 2016 (Figure 6.2).

Airports in Japan

Narita International Airport is situated approximately 60 km east of the city centre in Tokyo and served over 30 million passengers in 2013, making it Japan's primary airport for international traffic. JACB estimated that Narita represented over 50 percent of international passenger traffic into Japan and 60 percent of cargo. The airport is served by 87 airlines to 143 non-stop destinations, the majority of which (70) are in the Asia-Pacific region (Figure 6.3).

To face the challenge of managing its capacity with the increasing international demand, it has been necessary for Narita to expand the number of available landing slots and adopt a more efficient air traffic control system. In keeping with the overarching national intent to strengthen Japan's position as a major hub in the region, Narita International Airport intends to capitalise on widening liberalisation through the broader national adoption of open sky policies, and a focus on attracting and expanding its LCC operations.

Kansai International Airport benefits from being located in one of Japan's major economic regions with a population of 21 million. Economic activity in the Kansai

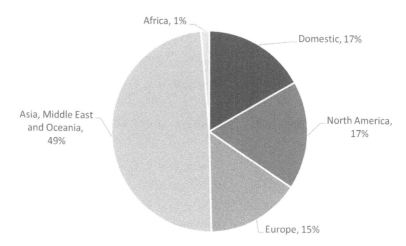

FIGURE 6.3 Destinations from Narita, 2019.
Source: centreforaviation.com and narita-airport.jp

region accounts for almost 1/5th of Japan's GDP. The tourism appeal of the Kansai region created by the extensive tourist attractions makes the area an attractive destination for international visitors. The development of the economic centre surrounding Kansai International Airport therefore supplements its integral role as an international gateway in Japan. In 2012, Kansai International Airport welcomed over 16 million passengers. In 2019, 76 airlines served the airport connecting to 88 destinations, the majority of which (58) were in the Asia-Pacific. Similarly, Chubu Centrair International Airport which opened in 2005 is also located in a major economic centre in Japan with a population of 22 million. Like Kansai, this area also accounts for almost 1/5 of Japan's GDP. The localisation of various international companies and manufacturing bases in the region has created sustained passenger and cargo growth opportunities providing strong evidence to support the construction of Centrair. Of these three major airports serving international traffic, Centrair is the least globally connected serving 53 destinations of which 30 are in the Asia-Pacific region.

Japan's air transport policy and its emerging views towards liberalisation

While Japan has increasingly pursued less restrictive policies in its bilateral air transport negotiations, Japanese officials initially maintained that the American approach to liberalisation cannot be applied wholesale in the Japanese context. Consequently, Japan has had a more modest perspective of liberalisation which has principally been motivated by capacity constraints due to slot availability and the relatively lower competitiveness of Japanese carriers (Oum and Chunyan 2019). Japan's progress towards a more liberal environment has been slow with one of its first liberal agreements signed with South Korea in 2007. The agreement allowed unlimited access to any international airport within Japan with the exception of Haneda and Narita where capacity constraints remained an issue.

Japan's softening approach to liberalisation is also reflected in its negotiations between Japan and the USA, resulting in an open skies agreement in late 2009 after decades of discourse. In the case of the USA, in the early 1990s, prior to technological advancements, Japan offered critical strategic positioning as a hub for US carriers to feed traffic into Asia. The lower competitiveness of Japanese carriers relative to their US counterparts led this to being a contentious issue with Japan arguing that US carriers would be more than capable of combining traffic from various US cities in Tokyo to feed their fifth freedom services, which would be to the detriment of Japanese airlines. In the case of the USA–Japan open skies agreement, negotiations ultimately resulted in US airlines being able to serve any point in Japan but with restrictions on the number of services to the two airports in Tokyo.

In attempts to support economic growth, Japan's air transport policy became closely tied to its economic strategy thus creating an even more enabling environment for further liberalisation. This refined policy approach has also supported the growth of Japanese carriers. The shift in approach to economic regulation of Japan's aviation industry is viewed as essential to spur the nation's economic growth and has led to greater thrusts towards open skies. This shifting perspective has been deemed consequential particularly since cost advantages have not been realised to make the airports more competitive regionally as hub airports. As further proof of the expanding liberalism in Japan's air transport policy, in 2009, Japan and the EU entered into a horizontal agreement to expand cooperation in air services.

The government agency responsible for international air transport policies

The Japanese air transport industry falls under the remit of the Ministry of Land, Infrastructure Transportation, and Tourism (previously Ministry of Transportation). The Japan Civil Aviation Board (JCAB) is the division therein responsible for civil aviation. The JCAB was not precluded from the impacts of emerging interests in liberalisation in the USA in the late 1970s. Negotiations between the USA and Japan during this period with respect to air cargo expanded the number of operators in the city pair which also encouraged Japan to adopt greater liberalisation on domestic and international fronts to increase competitiveness (Alexander, 1996). The JACB had advocated that with an increasing thrust by the Japanese government to promote Japan as a tourism destination globally and with the rapid expected expansion of traffic in the Asian region, Japan's aviation sector is strategically positioned to play an integral role in the ASEAN region. To this end, an expansion plan was implemented to facilitate the growth of its major international airports including Narita, Kansai, and Chubu Centrair.

The Civil Aviation Bureau comprises the Air Navigation Services Department, the Aviation Safety and Security Department, Aviation Network Department, and the International Affairs/Aviation division (Japan Civil Aviation Bureau, 2008). Specifically, international air transport affairs are managed by the Planning Division within International Affairs. The division is broadly responsibly for establishing policies with respect to international service agreements, developing the wider aviation network and the administration of airports. Surveillance of international air carrier operations is overseen by a separate division—the Aviation Industries division. There is also general oversight of JACBs overarching policies with respect to its strategic developments through the Aviation Strategy Division. Japan's approach towards air transport is therefore executed through the combination of these three divisions. The Aviation Strategies division can be perceived as shaping Japan's approach to liberalisation and international air transport in keeping with perspectives on aviation in relation to national economic and development. The direction developed between the Aviation Strategies and International Affairs divisions are executed by the Planning division.

China–Japan bilateral air transport relations

Air transport services between China and Japan started in 1974 with one carrier designated by each side. The ASA was signed in conjunction with an agreement on trade and was a consequence of the attempts at reconciliation. Similar to how Taiwan (part of PR China, hereinafter referred to as Taiwan) has been at the centre of various forms of geopolitical diplomatic contentions, consensus on air transport relations between China and Japan also centred on matters surrounding Taiwan during the negotiations. The joint communique between the two nations in 1972 eventually gave rise to the "Treaty of Peace and Friendship" which was signed in 1978. Hsiao (1977) recapped that in 1973, Chinese officials expressed concern over Japan's approach to Taiwan's China Airlines (CAL), which, in the opinion of the PRC, only emboldened the perception that Taiwan was more than a province of the PRC. As testimony, China contended that Japan should, *inter alia*, cancel Taiwan's CAL's routes between Taiwan and Japan—a request which Japan opposed.

An air services agreement was subsequently reached between the nations in April and initialled on 24 May 1974 with Japan Airlines (JAL) and CAAC being designated to serve points into and beyond China and Japan, respectively. JAL was designated to serve Beijing and/or Shanghai with CAAC receiving rights to serve Tokyo and/or Osaka. JAL was also permitted to take traffic beyond China to London via intermediate points including New Delhi and Mumbai in India; Karachi, Teheran, Beirut, Cairo, or Istanbul; Athens or Rome; and via Paris. CAAC was similarly permitted to take traffic beyond Japan via intermediate points including Vancouver or Ottawa in Canada, and four points in Latin America.

Discussions in the 1980s

In 1986, the Japanese government initiated a policy change in international air transport, advocating multiple designation and more destinations to be served. The change in attitude brought about seven Chinese carriers operating 383 passenger and 44 weekly cargo flights from 18 Chinese cities to 18 Japanese destinations and three Japanese carriers operating 210 passenger and 46 cargo frequencies per week from five Japanese cities to nine Chinese destinations in 2010 (Figure 6.4). In 2006, of the 51.53 million international passengers originating from Japan, 10.8 million travelled to China, accounting for almost one-fifth of the country's total. A China–Japan Air Transport Dialogue was established in the same year too, which holds meetings at a ministerial level every year, with the primary objective being to encourage and promote collaboration in areas such as safety, investment, air traffic control, and aviation security between the two countries. A liberalized air transport services arrangement was one of the key issues on the agenda at each Ministerial Dialogue, with progress being made in removing restrictions on bilateral operations, though slowly. After the dialogue, Chinese carriers have been more aggressive in their expansion into Japan than Japanese carriers. This grew from 7 to 12 designated Chinese airlines servicing 17 Japanese cities, while the 3 Japanese carriers serviced 10 Chinese cities (Figure 6.4).

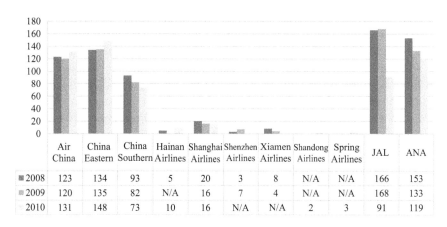

	Air China	China Eastern	China Southern	Hainan Airlines	Shanghai Airlines	Shenzhen Airlines	Xiamen Airlines	Shandong Airlines	Spring Airlines	JAL	ANA
2008	123	134	93	5	20	3	8	N/A	N/A	166	153
2009	120	135	82	N/A	16	7	4	N/A	N/A	168	133
2010	131	148	73	10	16	N/A	N/A	2	3	91	119

FIGURE 6.4 Comparison of weekly frequencies of passenger operations between China and Japan from 2008 to 2010.

Source: www.caac.gov.cn.

In the same vein, Sino-Japanese political relations also have negative implications for economic trade. It was reported that in 2012, China Eastern Airlines witnessed an 18 percent drop in sales during the month of September fuelled by public political discontent between the two nations. Similarly, Japanese airlines reported cancellations on 60,000 tickets over a two-month period during the public spat (China Economic Review, 2012). Just one year prior, in 2011, Japan and Taiwan neared completion of the negotiation of an open skies agreement to liberalise air transport arrangements between the two nations. By 2012, when the agreement had been concluded, the carriers were able to operate unlimited passenger and cargo services. The agreement facilitated free movement largely removing government controls with the exception of Japanese airports Narita and Haneda and Chinese airports at Beijing and Shanghai. While the secondary airports were growing in appeal due to liberalisation, Haneda was also experiencing increasing capacity into China which led to the approval for an increase in services between Shanghai and Beijing and Haneda and Guangzhou. OAG data suggest that between 2012 and 2015, flights increased by 38 percent and airline seats by 30 percent.

The growth in traffic between the two nations is substantiated by the close proximity and economic interdependence. Gradual improvements to the bilateral relations have resulted in a continuous rise in air traffic. With Chinese visitors being one of the primary source markets for Japan, Chinese carriers have continued to outpace their Japanese counterparts in their market access and growth in the market. In 2015, Japan witnessed a rapid spike in arrivals from Chinese visitors at over 100 percent year-on-year growth moving from 2.9 million visitors to over 4.9 million. While Japanese tourism growth in 2015 was sustained across various source markets, China represented significant year-on-year growth in the China–Japan tourism market. This was attributed in part to a weaker yen which made travel to Japan more appealing, the general easing of travel restrictions and, improved relations between the two nations. The year 2015 also had marked the 70th anniversary of the end of World War II and the passing of the commemoration without aggravating ongoing disputes avoided stifling the burgeoning Chinese tourism demand for Japan. Just a year ago in 2014, the leaders of the two nations had once again reconvened for their first formal meeting in over two years. With relations between Japan and China set on a gradual path of improvement, by 2018, arrivals from Chinese visitors had grown to 8.38 million, establishing a new record over the previous year's high of 7.35 million. Conversely, Japanese outbound tourism and visitor arrivals to China are comparably less significant. In 2006, a total of 3.7 million Japanese visited mainland China whereas by 2015, this had fallen to 2.4 million according to statistics from the Japan National Tourism Agency as detailed in Figure 6.5. Japanese visitor arrivals to Taiwan were recorded at 1.1 million in 2006 rising to 1.89 million a decade later.

Cognizant of the sustained interest in Japan from Chinese visitors, Japan Airlines and China Eastern sought permission to establish a joint venture. This

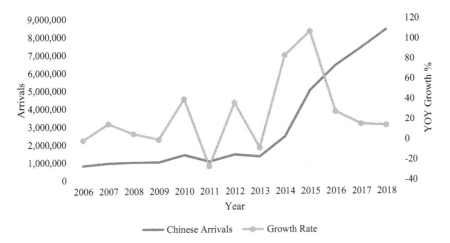

FIGURE 6.5 Chinese visitor arrivals to Japan, 2006–2018.
Source: statistics.jnto.go.jp.

expanded their previously established codeshare agreement which dates to 2002. With approval, the joint venture will also extend the respective carrier's networks into the partner's domestic markets. It could provide China Eastern with access to 50 cities in Japan whilst complementing Japan Airline's network with 80 domestic Chinese routes (Ch-aviation, 2018). The two airlines reportedly carry 30 percent of capacity on the country pair with China Eastern serving as the overall market leader covering 21 percent of the market share (Ch-aviation, 2018). Prior to 2015, ANA had the majority capacity share on the country pair followed by China Eastern; however, the year 2015 proved to be a pivotal turning point as evinced in the growth in Chinese arrivals in Japan.

In the two years prior to this shift in market dynamics, there were 5.8 million seats in 2013 followed by 6.7 million in 2014. In 2014, All Nippon Airways had been the market leader on the country pair with 1.7 million seats followed by China Eastern which served the market with 1.35 million. When China Eastern replaced All Nippon as the market leader in 2015, total seat capacity had risen to 8.8 million up from the 6.7 million in 2014. The 2.1 million additional seats were principally driven by an increase in capacity from Chinese carriers. While Japan Airlines' seat capacity declined by 83,000 in 2015, seat capacity by All Nippon Airways increased by an almost equitable 80,506. Growth on the country pair was fuelled by the Chinese carriers, led by China Eastern which supplemented the market by introducing over 631,000 seats. This equated to half of the carrier's total seat capacity on the country pair in 2014. Of the 2.1 million additional seats which were introduced into the market in 2015, 1.5 million were attributed to increases by the top four Chinese carriers on the country pair, namely China Eastern, Air China, China Southern, and Spring Airlines (Table 6.1).

TABLE 6.1 Top 4 Chinese carriers capacity growth

Top 4 Chinese carriers capacity growth (2014–2015)			
	2014 Capacity	2015 Capacity	Capacity growth
China Eastern	1,357,536	1,988,816	631,280
Air China	1,313,895	1,551,777	237,882
China Southern	497,146	720,807	223,661
Spring Airlines	238,544	659,160	420,616
Total			1,513,439

Source: centreforaviation.com.

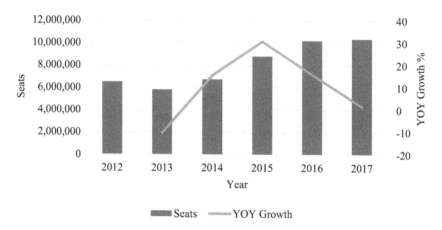

FIGURE 6.6 China–Japan seat capacity, 2012–2017.
Source: www.caac.gov.cn.

The subsequent year in 2016, total seat capacity had grown by a further 1.4 million to a total of over ten million seats as depicted in Figure 6.6. Strongest capacity growth during this period was precipitated by All Nippon Airways which augmented the market with a further 273,503 seats, followed by China Eastern which added 166,218. The top six carriers on the country pair (China Eastern, ANA, Air China, Japan Airlines, China Southern Airlines, and Spring Airlines) were responsible for 51 percent of the additional capacity (Figure 6.6).

China and the Republic of Korea (South Korea)

Situated between China and Japan, and fought over by the USA and Russia, the Korea Peninsula has long played a central role in Asia's geopolitical affairs (Pan, 2006). The half-a-century-long USA–South Korea political and security alliance established after the Korean War blocked any contacts between China and South Korea, which only changed in the early 1990s. With economic growth since the 1980s, South Korea has shifted its focus towards increasing regional ties with its Asian neighbours including China, with whom diplomatic ties had not been established until 1992. South Korea's position had been that by developing itself into a regional logistics and business hub, it would be able to play a bigger role in balancing the various parties in the region (Pan, 2006). Being influenced by Chinese culture and civilisation, the Korean public also believe that in the long term, China would serve to be a more important ally to them than the USA, which can be used to counteract the American influence (Roy, 2004).

China also shifted its Korean policy in the early 1990s, preferring to cultivate a relationship with the South Korean government that would have much more to offer economically (Roy, 2004). China was more motivated to attract investment and technology from South Korea whose economy had started to grow in the 1980s and soon became one of the Four Tigers in Asia. The two countries agreed politically to jointly oppose the Japanese Prime Minister's visit to the Yasukuni war shrine as well as working together to deal with North Korea's nuclear ambition. The improved political relationship stimulated bilateral trade, with China overtaking the USA to become South Korea's largest trade partner in 2003 and South Korea being China's fifth largest foreign investor. In 2004, the bilateral trade reached $90 billion, a 42 percent increase from the previous year.

Since the 1990s, South Korea has maintained a trade surplus with China that has largely been driven by exports including electrical machinery, chemicals, and plastics. By 2014, South Korea had still been able to maintain this trade surplus. Most of its exports included electrical machinery, television parts and accessories, and similar equipment which accounted for over 35 percent of its exports to China, valued at $51 billion. Medical or surgical parts and accessories, photographic, and cinematographic parts and accessories accounted for 15 percent of the trade between the two nations and was valued at $21 billion. In 2014, South Korea exported over $145.3 billion in goods and services to China. Chinese exports to South Korea were comparably smaller at a value of only $90.1 billion. Similarly, China's primary exports to South Korea also included items such as electrical machinery, television parts, and accessories and similar equipment (Worldbank Trade Database, n.d). The extensive trade between China and South Korea gave rise to a China–South Korea Free Trade Agreement signed in 2015, which encompassed almost $300 billion in goods and services traded between the two (Schott et al., 2015). However, Schott et al. (2015) observed that the agreement failed to meet the expectations of the business community by overlooking opportunities for further liberalisation in attempts to bolster political objectives. They contended that it would thus implicate attempts at economic integration in the case of the China–Japan–South Korea FTA.

South Korea and its air transport industry

Air transport in South Korea began to increase in the late 1980s and early 1990s in concert with the expanded air transport operations provided by the entrance of new domestic carriers such as Asiana and the broader economic growth fuelled by the five-year development plans. Relaxations on international travel restrictions also began to emerge during this period supporting the revised policy approach to adopting a greater degree of liberalism in air transport policy. Prestigious sports events such as the 1988 Seoul Olympics served as a pivotal turning point in South Korea's history where the true potential of international passenger traffic growth became most pronounced. Consequently, the government invested in developing a new contemporary international airport, recognising that the existing airport capacity would constrain the nation's ability to benefit from international arrivals.

Of the 15 airports across South Korea, Incheon International Airport serves as the biggest hub with the majority of international flights being concentrated here. International air services are available at seven other airports including Gimpo, Jeju, Gimhae, Cheongju, Daegu, Yangyang, and Muan. Sustainable economic growth in South Korea has attributed to continuous air traffic growth opportunities, and after the relaxation of its protectionist measures, domestic operations have experienced more growth. Korean airports witnessed over 42 million passenger movements in 2011 positioning it as the sixth largest country by air transport volume (koca.go.kr) and third by freight. It also serves as a major transhipment point for international cargo with an average of 3.4 million tonnes being transported from Korea to other international points. In total, Korean airports handle approximately four million tonnes of cargo (International Transport Forum, 2019). Consequently, Korea's growth has been sustained through its adoption of some of the most liberal approaches to air transport agreements relative to other competing Asian counterparts such as Japan and China.

Incheon was conceptualised and opened in 2001 to serve as Korea's major hub and to capitalise on the potential air transport growth. Its strategic position with close proximity to China's north-eastern region also augments South Korea's overall ability to capitalise on growth potential from China as a market. In keeping with this growth, Incheon International Airport began to position itself to serve as a hub in Northeast Asia benefiting from its geographical positioning. Incheon's ability to extend its hub status has not been without the sustained support and vision of the Korean government and its policies to international air transport development. It was developed in multiple phases each tailored at reinforcing the airport's position as the premier hub in Northeast Asia.

The prudence of the investment was evident in the airport's consistent leadership position on a global scale in airport service quality. Copping this prized title over a span of a decade from 2005 to 2013, the airport has brought significant international attention to South Korea. By 2017, the airport had undergone three phases of development which expanded its passenger terminal and cargo facilities, added additional runways and improved connectivity between terminals. Overall, Incheon International Airport has maintained that its primary goal in its developmental phases has been to continuously enhance the airport's capacity as the logistics hub of Northeast Asia and a premier provider of airport service quality in the scope of facilities available. Incheon's globally competitive positioning has also supported its prevalence relative to other airports globally representing the 15th largest airport by passenger traffic (Kin et al., 2015).

Its growth as a hub has also been ensured through the revised policy approaches to liberalise air transport and encourage greater competition among Korean carriers. Between 1988 and 1997, more international routes were awarded and Asiana Airlines' ability to enter into the domestic market enhanced competitiveness, ending years of Korean Air protectionism creating greater airlift and air travel demand. Then between 1988 and 2003, licensing requirements were revised enabling new carriers to emerge giving rise to less non-scheduled services. These developments occurred in concert with the wider and more rigorous development of air service agreements. As a result of this more liberal policy stance, by 2013, air transport accords and agreements had been negotiated with over 93 states. Of these agreements, multiple designations are permitted for less than 85 states- a vast contrast to the initial nascent position promoting Korean Air protectionism.

South Korea's policy on international air transport

South Korea's air transport policy position was largely influenced by the relationship between the privatised Korean Airlines, the Civil Aviation Board, and the Ministry of Foreign Affairs. Prior to 1963, the Korean aviation industry was managed by a division under the Ministry of Transport. The Civil Aviation Board (KCAB) was then established as a specialised division to effectively promote and develop South Korea's aviation industry. In 1969, Korean Airlines (KAL) was returned to private sector ownership in an attempt to improve its financial fortunes after enduring persistent losses. Hong (1989) describes the Korean government as having been heavily involved in the management of KAL prior to its privatisation which also constrained its viability. The privatisation of the carrier did not entirely decouple this relationship. Instead, it perhaps reversed it, creating a private sector influence on South Korea's air transport policy. Hong (1989) chronicled the evolution of the industry showcasing a dependence on KAL by the KCAB due to its miniscule staff compliment and comparative lack of expertise. KALs international network and global experience created a mismatch in the human resource capacity between itself and KCAB, which reportedly made it difficult for KCAB officials to effectively counter positions by KAL. According to Hong, this dynamic was so consequential that officials within the KCAB sometimes had to solicit the guidance of KAL employees to formulate the policy positions to be recommended to their superiors.

As a consequence of this interwoven relationship, KAL was able to shape Korean Air Transport policy by encouraging the market access it required to develop its network. In some way it could be argued that this created an environment of protectionism, not so much because of the government's intent to preserve growth for KAL but more so because of KCAB's inadequacies. The involvement of the Ministry of Foreign Affairs in negotiating the agreements was also more procedural than substantive to the extent that Hong reports the MOFA also took its advice unabridged from KAL and KCAB because it too did not have designated/ specialist aviation officers. The market was characterised by this scope of protectionist policies up until 1988. At this point Korean Air's reign as the sole designated air carrier was dissolved when the government approved the operations of Asiana to offer domestic services.

Relaxing controls on market access were not immediate though. The restrictive nature of the environment was prominently portrayed in the government's continued inclination to solely designate Korean Airlines on long-haul international services while offering some shorter routes to Asiana (International Transport Forum, 2019). Since 1990, the continuous reduction in restrictions showcased a more progressive approach which has encompassed changes such as reductions in licensing requirements and specifications in order for carriers to fly international routes. Initially, Korean carriers were required to serve a probationary period of two years serving domestic routes with 20,000 air traffic movements before being permitted to fly some international services. The initial probationary policy constrained the ability of domestic carriers to develop their international competitiveness. Consequently, while economic development plans were ongoing to propel economic growth, domestic constraints on carriers implicated their ability to benefit from, or further propel these economic opportunities. The Aviation Act enacted in 1961 indoctrinated these initial protectionist measures. By 2012, the Act had been effectively amended some 61 times through partial amendments, a complete revision, and through amendments by other related acts and legislation. The process of unbundling the protectionism entrenched through the Aviation Act created the opportunities for greater competition. This was done in part by increasing the potential for domestic carriers to serve international routes. This increased liberalism supports the overarching attempt to develop its airports as stronger hubs and spoke operations and to rival their regional competitors as vibrant and effective hubs. Developing as an effective hub necessitates carriers having less restrictions in access and frequencies, and therefore, South Korea has actively pursued more bilateral and multilateral agreements.

The government agency responsible for international air transport and bilateral negotiations

In an effort to emerge from its initial protectionist approach to the industry, the agency responsible for oversight of Korea's air transport industry underwent various transformations. In 1994, the transformation attempted to reduce government intervention firstly by centralising the organisations with the oversight of the industry. Six divisions were created under the then titled Ministry of Construction and Transport to create a more effective approach to managing the nation's overall infrastructure. The divisions were again restructured in 1999 to further centralise operations by creating less divisions with branch offices at regional airports. Extending branch offices into regional airports reinforced the extent to which there was an increasing focus on supporting domestic competitiveness. By 2008, the Ministry of Construction and Transport was reorganised and retitled as the Ministry of Land, Transport, and Maritime Affairs in concert with a sustained quest to expand economic development potential by centralising the nation's pillars of economic infrastructure and in 2009, the industry was refashioned to more closely align with its current construct where the Aviation Policy Office was created to further centralise air transport policy development and execution.

South Korea's policy on civil aviation is now administered through the Ministry of Land, Infrastructure, and Transport where the civil aviation office is constituted. The office comprises several core bureaus overseen by a Director General (DG) for Aviation Policy, Director General for Aviation Safety Policy, and a Director General for Airport and Air Navigation Policy, respectively (koca.go.kr). Generally, the formulation and advancement of South Korea's policies regarding liberalisation, international air transport policy, and aviation security fall under the remit of the DG for Aviation Policy. The DG for Aviation Safety administers the air transport licenses, ensures compliance with flight standards, and issues airworthiness certificates. The DG for Airport and Air Navigation Policy develops and administers airport development policy and planning, and oversees air navigation facilities and safety protocols at Korean airports.

More specifically, the Aviation Policy Bureau is primarily involved in decisions with respect to international air transport and South Korea's bilateral negotiations. The Bureau comprises various divisions, including Aviation Policy, the International Air Transport, Aviation Industry, Aviation Security, and ANS Standards and Oversight. Aviation Policy administers major mid- and long-term policy developments in keeping up with the overall mandate of the Civil Aviation Office as advocated by the Ministry of Land, Infrastructure, and Transport. Stakeholder rights are also overseen by this division and matters concerning domestic licensing and operations. The International Air Transport Division works specifically to develop and negotiate ASAs and similar aviation accords and matters related to the international routes. While policies related to domestic fares are prescribed under the Aviation Act, international fare policies are dictated under the respective ASAs and as such matters concerning fares are also under the remit of the International Air Transport division. The division's execution of its mandate is developed in concert with the policies set forth by the Aviation Policy division.

Similarly, the aviation industry division monitors and develops plans for the industry related to technology, issues concerning maintenance and various mid- to long-term development plans. One of the most critical functions the division plays in respect of the aviation accords encompasses its experiences regarding aviation business/ commerce-related laws. Its insights are particularly prudent in support of the ongoing efforts to liberalise South Korea's air transport sector to ensure fairness and competitiveness to sustain the industry's growth potential. Two other divisions within the bureau are the Aviation Safety which enacts aviation laws and administers airport security plans and ensures the industry's overall compliance with international security standards. Similarly, the ANS Standards and Oversight Division also has a safety oversight role which is specifically related to air navigation and air traffic control.

China–South Korea bilateral air transport relations

China–South Korea air transport services began in 1990, two years prior to the official establishment of diplomatic ties. They started as a charter service in order to satisfy the traffic needs for the 1990 Asian Games, which was later converted to scheduled operation when the bilateral ASA was signed in October 1994 in Seoul. Initially, the two designated carriers, Korean Air (KAL) and Asiana Airlines only operated to Beijing, Shanghai, and Guangzhou. As Korean investment in China expanded from the east coast in Shandong, Beijing, Tianjin, Liaoning, and Heilongjiang to cities in central and western China, Korea's air transport services were also extended to destinations such as Xi'an, Wuhan, Kunming, Changsha, and Zhengzhou. According to CAAC data, in 2010, eight Chinese carriers operated from 26 cities to 7 Korean destinations with 377 weekly frequencies for passenger services and three Chinese all-cargo carriers operated from five Chinese cities to Seoul with 24 weekly frequencies. From the Korean side, the two Korean carriers originated from 5 cities to 25 Chinese destinations with 330 weekly frequencies for passenger services, and two Korean cities to six Chinese destinations with 43 weekly frequencies for all cargo services (Figures 6.7 and 6.8).

South Korea remains fully aware that airports in Northeast Asian countries are competing fiercely for air transport demand, especially cargo, which are expected to take up to 42.2 percent of the world market in 2025 (Korean Ministry of Land Infrastructure and Transport, 2017). Airport expansions in Beijing, Shanghai, Guangzhou, Incheon, Nagoya, and Tokyo Haneda are just demonstrations of such competition. To strengthen Korea's regional hub status, the government has set the goal of preparing a foundation that would allow safe and convenient air travel and at the same time develop its aviation industry into a leading one in the Asia-Pacific region (Korean Ministry of Land, Infrastructure and Transport 2017). To this end, the Korean government accelerated the pace of liberalising its air transport market, with 29 open skies agreements concluded by 2008 compared with 11 in 2002. A further agreement was signed between

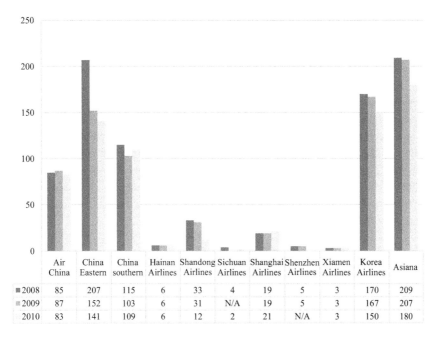

	Air China	China Eastern	China southern	Hainan Airlines	Shandong Airlines	Sichuan Airlines	Shanghai Airlines	Shenzhen Airlines	Xiamen Airlines	Korea Airlines	Asiana
2008	85	207	115	6	33	4	19	5	3	170	209
2009	87	152	103	6	31	N/A	19	5	3	167	207
2010	83	141	109	6	12	2	21	N/A	3	150	180

FIGURE 6.7   Weekly frequencies of passenger operations between China and South Korea.

Source: www.caac.gov.cn.

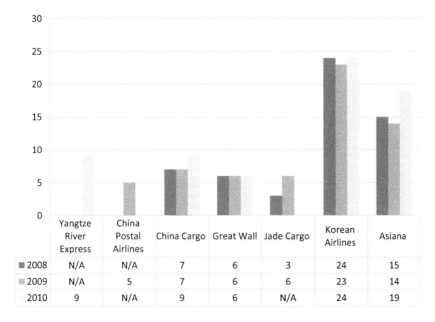

	Yangtze River Express	China Postal Airlines	China Cargo	Great Wall	Jade Cargo	Korean Airlines	Asiana
2008	N/A	N/A	7	6	3	24	15
2009	N/A	5	7	6	6	23	14
2010	9	N/A	9	6	N/A	24	19

FIGURE 6.8   Weekly frequencies of cargo operations between China and South Korea.

Source: www.caac.gov.cn.

the two nations in 2004 to expand access by increasing frequencies, removing restrictions on designation, and expanding fifth freedom rights.

South Korea pushed hard for a regional liberalised air transport market and proposed to launch a China–South Korea Air Transport Cooperation Forum in August 2005, with the second Cooperation Forum held in December 2007. The benefit of the forums between the duo cannot be understated. In 2006 further liberalisation was effected in the Korea–Shandong and Hainan markets and China even proposed full liberalisation across the country pair by 2010. The position of the Chinese government regarding full liberalisation was subsequently tempered by more moderately liberal policy perspectives driven by external and internal pressures such as fears that Chinese carriers were not equitably competitive (Lee, 2014). A liberalised air transport zone for the region is one of the key items of the Cooperation, although progress is slow.

References

Alexander A., (1996), *Domestic aviation in Japan: responding to market forces amid regulatory constraints*, Center on Japanese Economy and Business, New York.

Armstrong S., (2012), The politics of Japan- China trade and the role of the world trade system. *The World Economy*, 35(9), pp.1102–1120.

Ch aviation (2018), JAL-China Eastern to form a joint venture, available at www.ch-aviation.com/portal/news/69392-jal-china-eastern-to-form-a-joint-venture

Chan S., and Kuo C., (2005), Trilateral trade relations among China, Japan and South Korea: challenges and prospects of regional economic integration. *East Asia*, 22(1), 33–50.

China Economic Review, (2012), Turf Versus Trade: China and Jappan cant afford a heated economic dispute amid their cooling economies, available at http://www.freetradedoesntwork.com/research/CER-1211.pdf

Chung C., (2014), *Contentious integration: post-cold war Japan-China relations in the Asia-Pacific*, Routledge, Oxon.

Hong S., (1989), Aviation policy making in South Korea, (Doctoral Dissertation), George Washington University.

Hsiao, Gene T., (1977), *The foreign trade of China: policy, law, and practice*, University of California Press, Berkeley.

International Transport Forum, (2019), Liberalisation of Air Transport, ITF Research Reports, Paris.

Japan Civil Aviation Bureau, (2008), Organization and function on civil aviation at ministry of land, infrastructure and transport, available at www.mlit.go.jp/koku/15_hf_000031.html

Kin J. K., Park J., Shim G., and Yoo H., (2015), Significance of Korea's aviation organization development., Issue 2, The Korea Transport Institute, Sejong-si.

Korean Ministry of Land, Infrastructure and Transport, (2017), Sectoral policies: aviation, available at www.molit.go.kr/english/USR/WPGE0201/m_36861/DTL.jsp

Lee J., (2014), Regional liberalisation in Northeast Asia (China, South Korea, and Japan), in: Duval (edited), *Air Transport in the Asia Pacific*, (pp. 217–234), Routledge, Oxon.

Leelawat N., (2015), Disaster recovery and reconstruction following the 2011 Great East Japan Earthquake and Tsunami: a business process management perspective. *International Journal of Disaster Risk Science*, 6(3), 310–314.

Mofa, (2011), Joint Study Report for an FTA among China, Japan and Korea, available at www.mofa.go.jp/mofaj/press/release/24/3/pdfs/0330_10_01.pdf

Oum T., and Chunyan Y., (2019), *Shaping air transport in Asia Pacific*, Routledge, London..

Pan, E., (2006), South Korea's Ties with China, Japan, and the U.S.: Defining a New Role in a Dangerous Neighborhood, available at, https://www.cfr.org/backgrounder/south-koreas-ties-china-japan-and-us-defining-new-role-dangerous-neighborhood

Pekkanen, S. M., and Tsai K. S. (2005), Late liberalisers, comparative perspectives on Japan and China, in Pekkanen S. M., and Tsai K.S., (2005) (edited), Japan and China in the world political economy, Routledge, Abingdon and New York, pp. 11–28.

Purdie, E. (2019), Tracking GDP in PPP terms shows rapid rise of China and India, available at https://blogs.worldbank.org/opendata/tracking-gdp-ppp-terms-shows-rapid-rise-china-and-india

Roy D., (2004), Special Assessment: Asia's Bilateral Relations: China-South Korea Relations: Elder Brother wins over Younger Brother, *Asia-Pacific Centre for Security Studies,* (10), 1–7. Schott J., Jung E., and Cimino-Isaacs C., (2015), *An assessment of the Korea-China free trade agreement*, Peterson Institute for International Economics, pp. 15–24.

Stuart H., (2014), *China's foreign policy*, Polity Press, Oxford.

World Bank, (2012), *The Great East Japan Earthquake, Learning from Mega Disasters*, No. 79352, Washington: World Bank.

Worldbank Trade Database, (n,d.), available at https://wits.worldbank.org/CountryProfile/en/Country/CHN/Year/2017/TradeFlow/EXPIMP/Partner/by-country

Xing Y, (2008), Japan's unique economic relations with Chin, economic integration under political uncertainty, available at http://www.eai.nus.edu.sg/BB410.pdf

7

APPLYING THE MICRO-MACRO LINKAGE APPROACH (PART I)

Analysis of factors at international and regional levels

The global regime and its impact on policymaking

China only became a member state of the United Nations (UN) in 1971 with the primary impetus to gain international recognition to break out of its global isolation that had existed since the establishment of its government in 1949 (Kent, 1999). Although inexperienced in the early days on the global political stage, China recognised the significance of engaging itself in international institutions. Politically, such participation would enable the country to leverage its position in the international community for its own benefits while playing a key role in constraining the behaviour of other international actors so as to prevent any decisions adverse to its national interests and infringements on its internal governing capacity (Lampton, 2001). Economically, the international institutions serve as the best platform to enable China to capitalise on opportunities to gain access to international resources for aid, investment, technology, and information, which will further facilitate its export-oriented economic growth, to enhance its international status from both political and economic perspectives and to achieve its strategic goal of becoming a modernised power in the global political system.

To gain experience in the international political arena, China initially chose to join a couple of international institutions which are more technically oriented rather than politically focused, thus becoming a Contracting State of ICAO in November 1971. The following decades saw the country to join more international institutions such as the World Bank and the IMF, as its feather pluming. By the beginning of the 21st century, China had been involved in more than 100 international institutions, where it could seek to optimise its national interest and play a major role in shaping the world order (Kent, 2007).

China's accession to the WTO

The WTO has evolved out of the General Agreement on Tariffs and Trade (GATT) which dates back to 1947, whose overriding purpose is to facilitate free trade, promote economic development, and the well-being of society. As one of the intergovernmental institutions within the UN system, the WTO stipulates rules and regulations on international trade with the consensus being achieved through multilateral negotiations among member governments. The sets of these rules, signed by the bulk of the world's trading nations, are at the heart of the WTO system, which is binding for every member state. The member state is hence obliged to honour the agreements signed under the WTO regime and abide by its rules. Accordingly, the member state is also obliged to ensure its trade policies are non-discriminative, transparent, and within the agreed limits.

China's endeavour to join the WTO started in the mid-1980s. The marathon negotiations between the two parties in the following years resulted in China officially becoming a WTO member states in November 2001. China's accession was significantly important for both parties. WTO was pleased to see its multilateral rule-based trade system extended to China, a large, dynamic, and increasingly important trading nation, who was expected to play a key role in multilateral trade negotiations. From China's perspective, the accession was a great opportunity to underpin its international trade and investment strategy.

An official WTO membership required China to demonstrate its commitment to honouring all WTO rules. Where discrepancy exists, it required China to dismantle its protective trade regime to bring it in alignment with that of comparable WTO member states (Thomson, 2003). The biggest challenge for China, was to transform its opaque and black-box style rule making and arbitrary system into a more transparent, fair, and open policymaking process and a rule-based system.

To achieve this, China made three tiers of commitment. The first tier was the commitment to the spirit of the WTO, such as free trade, most favoured nation status, national treatment, and transparency, which is expounded in the various documents of the WTO. The second tier was the commitment to the set of rules governing business practice and trade for industries such as agriculture, information technology, banking and financial services, telecommunication, and transport and logistics. The third tier was the commitment made in the bilateral agreements which China has signed with its major trading partners, especially with the USA (Ching and Ching, 2003).

To this effect, China developed a Schedule of Specific Commitments (the Services Schedule), which it committed to achieving within a certain period of time. The Services Schedule identified several industry sectors to which

China would apply its WTO market-access and national-treatment obligations. It also justified the exceptions from these obligations which it intended to maintain (Yeaman, 2003). The Services Schedule also included commitment to replacing the incompatible laws and regulations with the revisions or elimination to ensure its full compliance with the WTO requirements (Kapp, 2002).

Regarding the transport and logistics sector, "Several Opinions on Enhancing the Development of a Modern Logistics Industry in China" (the Logistics Opinion), jointly issued by CAAC and several other ministries, became effective in July 2002. The Opinion welcomed foreign investment in the logistics sector to provide services for international distribution and third-party logistics (Yeaman, 2003).

Regarding civil aviation, the WTO only covers aircraft maintenance and computer reservation system (CRS). International air transport is treated as a trade in services under the WTO. Although endeavours have been put forward since the 1980s, at the Uruguay Round of Negotiations, to reach a unified regulatory regime governing the economic aspect of civil aviation activities, the rounds of these multilateral negotiations have yet to lead to any consensus to date. One of the contributing factors is that member states of ICAO are obliged to adhere to the Chicago Convention, which authorises individual member state's government to agree to transborder activities of international air transport in a bilateral basis. It is thus at the hands of an individual government to determine to what extent a liberalised transborder air transport market is to be pursued to allow free movement of people and goods.

The WTO's multilateral approach to liberalising cross border air transport activities, remains elusive until today. Nevertheless, in the Doha Round of Negotiations launched in 2001, WTO set up a working group to assess the impact of liberalisation of international air transport. The ultimate objective is to apply WTO's fundamental principles of non-discrimination, free trade, and liberalisation to international air transport.

To comply with WTO's rules with respect to air transport, China is committed to the following:

- In terms of aircraft maintenance, allowing Chinese aircraft to be repaired outside the country; allowing the establishment of joint-ventured maintenance facilities with the Chinese side taking the controlling stake and being eligible to provide services in the international market;
- In terms of CRS, allowing foreign CRS to connect with Chinese CRS and provide services to Chinese carriers and their agents; however, China would not allow foreign companies to invest and operate independent CRS services;
- In terms of pricing, China should not discriminate against any foreign persons, corporate, and foreign-invested businesses in the provision of air transport services;

ICAO and its call to liberalise international air transport

ICAO, as a specialised agency of the UN with 192 member states, is the international institution that stipulates rules and regulations governing international air transport activities. The regulatory regime focuses more on technical standards than commercial activities, as a consequence of failing to reach an agreement in Chicago in 1944.

> The aims and objectives of ICAO are to develop the principles and techniques of international air navigation and to foster the planning and development of international air transport so as to ensure the safe and orderly growth of international civil aviation throughout the world; to meet the needs of the people of the world for safe, regular, efficient and economical air transport; and to promote safety of flight in international air navigation.
>
> *(Article 44, Chicago Convention)*

Although ICAO is empowered with law-making authority, its policies and guidelines with respect to the economic aspect of international air transport activities (such as Assembly Resolutions, Declarations, and Recommendations adopted at air transport conferences) are generally of a recommendatory nature. This is in contrast with the Standards and Recommended Practice (SARPs) adopted by ICAO pertaining to the technical aspect of the aviation activities which are binding on its contracting states (Wang, 2011). Hence, it is at the discretion of the contracting states to determine whether to accept the policy guidelines and to further decide whether and how to implement it (Wang, 2011). Having said that it is noteworthy that as such a policy or guideline is adopted at a global level, in most cases by consensus, it carries the weight of international acceptance, making it a kind of "moral obligation" for ICAO contracting states to adopt, follow, or apply in their regulatory practices (Wang, 2011).

ICAO had been cautious in advocating liberalising international air transport market between its member states in the 1990s, the first decade after the USA officially launched the open skies initiatives in 1992. ICAO's view was that such a practice is, by nature, in contrast to the principles of the Chicago Convention, which upholds sovereignty and equal and fair opportunity for each and every member state. The aim of the Chicago Convention is "to ensure that the rights of Contracting States are fully respected and that every State has a fair opportunity to operate international airlines and to avoid discrimination between the Contracting States" (Chicago Convention,, 1944).

Its first conservative call came in 1994, two years after the USA signed its first ever open skies deal with the Netherlands. The "Declaration of Global Principles for the Liberalisation of International Air Transport in 1994" was issued at the Fourth World Air Transport Conference. Instead of urging its member states to take action in this respect, having "noted the changes in the regulatory and operating environment of international air transport brought about by economic development, globalization, liberalization and privatization", ICAO reiterated the

> critical importance of safety and security in international air transport and reaffirmed that the basic principles of sovereignty, fair and equal opportunity, and cooperation set out in the Chicago Convention should continue to provide the basis for future development of international civil aviation.
>
> *(Abeyratne, 2003)*

It further "called on the Member States to ensure a high level of safety and security and to respect national sovereignty" (Abeyratne, 2003).

Such a cautious and conservative approach towards liberalisation of international air transport apparently did not motivate nor encourage its member states to pursue a liberalised policy. The USA was left as the sole hegemon in the world to promote its beliefs in liberalising the industry.

The radical shift of ICAO's approach to liberalisation took place in March 2003, almost eight years after the 1994 Declaration, when the Fifth ICAO Air Transport Conference was convened in Montreal with 794 participants. ICAO, for the first time, declared its "full appreciation of globalization and liberalization, calling its Member States, to the extent feasible, to liberalise international air transport market access for carriers and their access to international capital markets and their freedom to conduct commercial activities" (ICAO Declaration, 2003). ICAO's view on liberalisation is that it is up to each member state to "determine its own path and pace of change in international air transport regulation, on the basis of equality and opportunity, using bilateral, sub-regional, regional and/or multilateral approaches according to circumstances." It reinforced the message that the objective of liberalisation is "to create an environment in which international air transport may develop and flourish in a stable, efficient and economical manner without compromising safety and security and while respecting social and labour standards" (ICAO Declaration, 2003). ICAO's declaration in 2003 did not oblige member states to take action on liberalising its aviation market. At best, it was seen as a push to raise the awareness of the member states of the irreversible trend of liberalisation.

As a Contracting State with a seat in the council, China honours its obligation under the Chicago Convention, upholds the principles of sovereignty, fair and equal opportunity, non-discrimination, interdependence, and cooperation and gives due regard to ICAO's policies and guidance in its decision-making with respect to international air transport (Wang, 2011). China follows the trend of liberalisation but decides on its own policy and pace of change. ICAO's update on the evolution of liberalisation of the air transport market made China aware of the current global trend. Its call for member state's action obliged China to reassess its approach to liberalisation to determine to what extent it needed to accelerate its liberalisation pace (Wang, 2011). As an immediate response, China declared at the same Conference that it would take an "active, progressive, orderly and safeguarded manner" in liberalising its air transport market (ICAO, 2013).

Impact of China's WTO membership and ICAO's advocacy for liberalisation on China's international air transport policymaking

International organisations are the institutional representations of internationalisation and interdependence (Kent, 2007). They represent an organising process of conflict management at the supranational level as well as a collective organising response to a multiplicity of control problems in a world of contradictory trends (Kim, 1994). The benefits of being involved in the international community not only include the assistance received to stabilise identities, codify interests, enshrine moral principles, and resolve cooperation problems, but also its commitment to complying with the norms, principles, rules, and their associated treaties (Kent, 2007). Thus, pressure to comply with the rules has already been built into the existing system that member states are obliged to take (Kent, 2007), as Chayes and Chayes (1995) rightly point out that international organisations constitute some of the primary sources of pressure for obtaining member states' compliance with regime norms. They and their treaty regimes not only encourage transparency, cut transaction costs, build capacity, and enhance settlement, but also, through a process of "jawboning," persuade parties to explore, redefine, and sometimes discover their own as well as mutual interests between member states.

China's accession to the WTO is regarded as one of the most significant landmarks in its process of restructuring the country's economy, a motive that has driven continuous and fundamental changes since 1978 (Yeaman, 2003). It is fully aware of its legal obligation to honour and comply with the WTO rules and act as a cooperative and compliant actor in the international regime (Kent, 2007). Being involved in international organisations and sitting in the Council have profoundly changed not only China's view of the world, but also its view of itself (Kent, 2007). China has accepted the obligations by acting responsibly to ratify the rules and associated treaties, which have been accordingly entrenched in its domestic legislation (Kent, 2007). "Once inside the WTO, China will strictly comply with the universally acknowledged market rules, implement open, transparent and equality-based policies of trade and investment, and endeavour to promote a multi-directional and multi-level opening-up in a wide range of areas", as China's President Jiang Zemin pledged (Kent, 2007). The ideology also started to change in the early 21st century when China became more integrated into the international community, which has brought about policy changes. China reduced the central control over its commercial policy, reduced its tariffs, removed non-tariff barriers and quotas, provided further protection of intellectual property rights, and eliminated many barriers to trade in goods and services (Kent, 2007). Barriers in conducting business have been iteratively relaxed so as to ensure that the pace of change is gradual, manageable, without a big-bang effect.

Although the WTO rules do not apply to economic activities in international air transport, some rules such as removing trade barriers apply to the air transport industry. Becoming a WTO member state has inspired China to take immediate and effective actions to overhaul its regulatory regime, ease market entry barriers and eliminate restrictions on operations for a fair and equal playing field for both domestic and international businesses. Specifically, the following policy changes were initiated with respect to air transport.

In August 2002, new regulations were introduced, allowing international investment in any existing public transport services, although the majority control of one shareholder was limited to a maximum of 25 percent. Statistics from CAAC showed that eight foreign invested projects were approved in 2002 with a contracted value of $80 million. The figure increased to $110 million in 2003 with six projects being approved (CAAC, 2004a). In early 2003, CAAC made a comprehensive and thorough reassessment of 224 regulations, orders and procedures that had been issued between 1980 and 2002, resulting in 129 being revoked or abolished. A second revision of its incumbent regulations was undertaken in 2005, resulting in the termination of another 40 regulations, thus having eliminated the restrictions on economic operations of the industry. Discriminatory airport and ATC charges against foreign carriers were adjusted with the new charges being introduced since 1 November 2005.

In the same year, a "Leading group to promote international air transport development" was set up, being responsible for developing the country's air transport development policy and its execution. A package of guidelines was thus issued, which included:

- Liberalising domestic air transport market in a gradual and staged manner;
- Opening up international freight market with immediate effect;
- Encouraging international air transport activities in central, western, and north-eastern regions;
- Strengthening gateways and hub airports;
- Developing an open, fair, and transparent mechanism to assess and approve international route licence applications;
- Reform of the approval procedures for additional international scheduled and charter operations;
- Encouraging and supporting airlines to open more international routes to destinations although there might be less market demand but support national strategic diplomatic objectives;
- Establishing and streamlining an effective mechanism to manage airport slots, CRS, and airport ground handling services; and
- Promoting simplified passenger travel procedures and e-tickets (CAAC, 2005).

Airport handling service in major mainland Chinese airports was further opened up for service providers, whose business bases were in Hong Kong SAR and Macau SAR, under the scheme of the mainland and Hong Kong Closer Economic Partnership Arrangement (CEPA), although such contracts would remain valid for a maximum of 20 years. Joint ventured ticketing agencies to distribute airline tickets were also allowed in the mainland (CAAC Annual Policy Statements of 2005, 2006 and 2008).

In response to ICAO's call to liberalise the air transport market, China reiterated its stance in its 2004 Annual Policy Statement, affirming that it would adjust its international air transport policy under ICAO's guidance (Wu, 2007). It would seek to establish air transport relationships with countries globally. In this pursuit, the Chinese government would abandon the protective notion, which only took into account the Chinese carriers' intention (Wu, 2007). Specifically, the policy statement included the following key objectives:

- To positively embrace the trend of air transport liberalisation;
- To support China's overall diplomatic objectives;
- To support China's Open-Up strategy and the objectives of social and economic development;
- To give due regard to the demand for international air transport to facilitate the country's foreign trade and tourism growth;
- To strengthen the air links between Western, Northeast, and Central China and the rest of the world;
- To balance the national interests, public interests, and industry interests;
- To develop the hub strategies;
- To enhance the overall competitiveness of the industry; and
- To seek cooperation to achieve a win–win situation, with an aim to encourage the expansion and growth of the whole industry which does not jeopardise China's national security.

To implement the policy, CAAC developed detailed strategies which were executed in each and every bilateral negotiation. In 2003, China agreed more relaxed arrangements with 12 countries at a bilateral level including those with the Netherlands, Australia, France, and Japan, all of which removed restrictions on designation (2003 CAAC Annual Policy Statement, 2004). In 2004, China conducted 15 bilateral ASA consultations including the USA, the UK, Germany, and Russia which significantly liberalised the market entry arrangements, although to a different extent. The 2004 Protocol with the USA, in particular, was hailed as a milestone in China's history in liberalising its international air transport market, which involved a comprehensive overhaul of the original restrictive bilateral ASA signed in 1980. In 2004 and 2006, respectively, China liberalised its arrangements with South Korea, with a full open skies arrangement being reached which applied to the South Korea–Shandong and South Korea–Hainan Provinces markets.

By May 2007, CAAC had achieved the following:

- Of the 108 bilateral ASAs signed between China and its counterparts, 74 allowed multiple designations, representing 68 percent of the total;
- Relaxed the control on route schedules with ten countries, allowing unlimited operation to intermediate, destination, and beyond points;
- Removed capacity restrictions on third and fourth traffic rights for both passenger and cargo operations with seven countries;
- Removed capacity restrictions on third and fourth traffic rights for cargo operations with six countries.

In addition to the above, China has allowed third-party codeshare arrangements with 34 countries, wet-lease arrangements with 22 countries, 5th freedom traffic rights for cargo operations with 7 countries and 7th freedom traffic rights for African airlines when operating cargo charters. China also agreed to relax its pricing regime to adopt the "origin principle" or the "double disapproval principle" (CAAC, 2007).

Summary

While sovereignty remains one of the key principles in the Chicago Convention governing international air transport, it has always been a paramount consideration for China when negotiating its international air transport agreements. Taking air transport as an integral part of its national defence system, China has taken a cautious and conservative approach towards liberalising the industry. However, its accession to international institutions such as WTO and ICAO has inspired the policymakers to adopt a more liberalised approach to international trade. ICAO's Declaration in 2003 has motivated the nation to take a proactive but staged and orderly approach towards liberalisation. For the first time, China issued an aviation White Paper in 2004, which reviewed China's policy change since 2003 and proposed an ambitious policy objective, with special focus on developing China's international air transport. CAAC, hence, developed detailed action plans, outlining the strategies and tactics to be adopted at each round of bilateral negotiations with its counterparts to optimise the outcome. The action plans served to achieve two objectives. On the one hand, it would facilitate to accelerate the liberalisation pace. On the other hand, it would also buy some time for its carriers to gain international experience. Acknowledging the irreversible trend towards liberalisation of air transport, the Chinese government was convinced, by then, that the most appropriate strategy was to take the challenge proactively rather than waiting for defeat. "We should open our market selectively and gradually in order to enhance market prosperity and improve the service level. We also need to enhance competitiveness of our businesses through introducing competition in an open environment", as CAAC's administrator declared (CAAC, 2007a).

The bilateral arrangements between China and South Korea, Japan, the USA, the Netherlands, and the UK that had been concluded before 2003 had various kinds of restrictions, though to a different extent, on market entry, capacity, pricing, and other doing business issues. After 2003, the arrangements agreed at each round of discussions contained significant relaxation of restrictions on designations, capacity, investment, and other business issues. Although what has been agreed with each country at each round of discussions is different, generally, market entry has been relaxed, constraints in route schedules have been eliminated, and capacity has been increased. The principle that the Chinese government upholds is to open the market cautiously, selectively, and gradually so as to create a relaxed market environment. Industry interest is no longer the exclusive concern of China's aviation authorities and institutions when formulating the objectives of bilateral negotiations with its counterparts, but rather, national interest and spill-over effects to other industries such as tourism and international trade have taken more weight in its considerations. Individual carriers are encouraged to participate in the competition for its survival.

Factors at the regional level and their impact on policymaking

China's link with regional institutions

ASEAN and its open skies initiatives

The Association of Southeast Asian Nations (ASEAN) was established on 8 August 1967 in Bangkok by Indonesia, Malaysia, the Philippines, Singapore, and Thailand with the signing of the ASEAN Declaration. Later joined by Brunei, Vietnam, Lao PDR, Myanmar, and Cambodia, ASEAN is now made up of ten member states in the region. As one of the world's most close-knit regional groupings (UNESCAP[1], 2010), ASEAN members recognise the economic potential to be derived through better collaboration to raise the living standards of the region's people. In contrast to the EU, which is formed on the basis of the Treaty of Rome with its laws, regulations, rules, and policy directives being binding to its member states, ASEAN was formed on the basis of an understanding contained in the declaration. After more than four decades of consolidating ties among its members, the "ASEAN Community" entered a new phase in December 2008 ratifying a legally binding charter that established the institutional framework necessary to propel the region's integration movement (UNESCAP, 2010).

Air transport has been recognised as an important sector for cooperation between ASEAN member states, given the role it plays in facilitating regional trade activities. In 1994, the ASEAN Plan of Action in Transport and Communications (1994–1996) was formulated to focus on the development of a multimodal transport system to improve interconnectivity within the region and to harmonise regulations (Yean, 2010). Effective management of air space was emphasised although no initiatives were taken to liberalise the markets. In the 1997's ASEAN Plan of Action in Transport and Communications, the development of a competitive air services policy was identified as a key area for the member states to collaborate. This led to the conclusion of the ASEAN Memorandum of Understanding (MOU) on Air Freight Services in 2002, although the contracting states were still limited to a maximum of 100 tonnes of carriage in freight operations per week. The momentum continued in 2008 when The ASEAN Multilateral Agreement on Air Services and the Multilateral Agreement on the Full Liberalisation of Air Freight Services was signed, which became effective in 2009 (Yean 2010). The ASEAN Single Aviation market was projected to be implemented in 2015. By early 2018, however, there were still existing constraints to be resolved including the integration of seventh freedoms and more comprehensive participation by member states such as Indonesia whose involvement had been limited to only five airports (Basu Das, 2018).

China's collaboration in air transport industry with ASEAN member states

China was first officially linked to ASEAN in 1991, when the then Chinese Foreign Minister Qian Qichen attended the opening session of the 24th ASEAN Ministerial Meeting (AMM) in Kuala Lumpur and expressed a strong interest in forging cooperation with ASEAN for mutual benefits. The relationship was formalised in July 1996, when China was accorded a full dialogue partner status at the 29th AMM in Jakarta.

Over the years, China has been proactively involved in ASEAN activities and pushed openly for the establishment of a China-ASEAN Free Trade Area (CAFTA). To ensure the successful launch of CAFTA, the then Chinese Premier Zhu Rongji proposed to start an annual Transport Minister Meeting (TMM) in November 2001 so as to enable the participants to discuss and coordinate policies and issues relating to transport collaboration and development, which is the key to facilitate the effective establishment of CAFTA. At the ASEAN Plus Three Summit[1] held in November 2001, China offered to open its market in some key sectors to ASEAN countries, five years before they reciprocated. China also offered to grant special preferential tariff treatment for some goods from the less developed states, such as Cambodia, Laos, and Myanmar (Sheng, 2003). Though the initial response was cautious, ASEAN accepted China's proposal with the Framework Agreement on Comprehensive Economic Cooperation being announced in November 2002 in Cambodia at the ASEAN-China Summit, which anticipated that 2010 would be the inception year for China and the six original ASEAN states to attend the Summit, namely Brunei, Indonesia, Malaysia, the Philippines, Singapore, and Thailand, while 2015 would be the year for the less developed ASEAN members of Cambodia, Laos, Myanmar, and Vietnam to join the summit (Sheng, 2003).

The year 2002 saw China's attendance at the first TMM, where the country's deputy administrator of CAAC delivered a speech on China-ASEAN air transport relations. In its second attendance at TMM in October 2003, China proposed a Memorandum of Understanding on Transport Cooperation, including air transport, calling for collaboration on infrastructure construction, project investment and financing, collaboration between airlines, exchange of information, and the establishment of an open skies zone through bilateral phased-in and progressive arrangements. The MOU set the blueprint for both parties in their pursuit for

> the relaxation of restrictions with respect to the number of points in route schedule, no limitation on 3rd, 4th, and 5th freedom traffic rights, no limitation on frequency and capacity as well as type of aircraft, and multiple airline designation.
>
> *(ASEAN, 2003)*

The MOU was signed in November 2004, which coincided with the conclusion of a multilateral agreement for passenger services reached between Singapore, Brunei, Thailand, and Malaysia (Yean, 2010).

China's championship in forging a free trade zone with ASEAN member states was rewarding. The country saw a 15 percent increase of trade year on year since 1995, and the trade volume reached $54.77 billion in 2002, which doubled to $105.9 billion in 2004. The share of China-ASEAN trade in China's total trade volume also increased from 7.0 percent in 1996 to 9.2 percent in 2004 (Ku, 2006). By 2013, trade had reached $443.6 billion and China had remained to be ASEAN's largest trading partner outside of ASEAN members. ASEAN member states represented China's third largest trading partner.

ASEAN countries have also concluded ADS (Approved Destination Status) arrangements, Lao PDR was the last of the members to be assigned ADS in 2005, with China with Malaysia, Singapore, the Philippines, Thailand, and Indonesia being among the top 20 tourist resource countries to China. China also would like to see the ASEAN countries to be a prospective market for its aircraft-manufacturing projects such as the ARJ21 (Williams, 2009). By forging a closer political relationship and providing magnificent economic support and concessions to these countries, China has gained the comparative advantages of enjoying both geographical proximity to and political membership of the ASEAN aviation environment (Williams, 2009).

Relations with peripheral neighbours Japan and South Korea

Forging close relations with its peripheral neighbours remained critical for China in strengthening its geopolitical position. Consequently, China had maintained close aviation relationships with ASEAN member countries along its periphery and in the wider region. This increased openness to the region was substantiated by the political reforms since the 1980s that catapulted its economic growth. The next frontier for this integration comes through further consolidation in relationships between China, Japan, and South Korea as three of the largest economies in the region. On 13 May 2012, the trio commenced preparatory talks with the intention of establishing a free trade agreement deepening their economic cooperation. The close relations between China and these two respective states discussed in Chapter 6 previously, signals the potential the agreement could offer for air transport in the region. However, progress on the FTA and potential air transport liberalisation among the three remain slow. Between 2013 and 2017, 12 rounds of negotiations had been held with frequent points of discussion surrounding trade and investment. Therefore, it remains to be seen the consequences for air transport relations under an enhanced liberalised framework for facilitating trade and investment across the three nations, given their size and relevance both in Asia-Pacific and the wider global economy.

In a broader sense though, China had been systematically reinforcing itself as a strategic partner in the face of what it had perceived to be growing western intervention, particularly in a region of economic importance to China. Therefore, in 2010, China and the ASEAN member countries established a free trade agreement (FTA) which to date has established China as the member grouping's largest trading partner, magnifying trade between the two partners. Attending this economic symbiosis enhanced aviation cooperation which could be reflected in the ASEAN–China Aviation Cooperation Framework. The framework was viewed as critical to support the comprehensive realisation of the China–ASEAN FTA by facilitating ease of movement of passengers and cargo. In part, the framework established a need to expand liberalisation of cargo and passenger services. In 2010, this gave rise to an ASEAN–China Air Transport Agreement which effectively created an open skies agreement removing capacity restrictions on third and fourth freedoms. Thus, capacity was principally only constrained by slot availability at the larger airports. In this instance, the more liberal market access facilitated expansionary growth for LCCs from SE Asia. The Chinese side however excluded liberalising services to Hong Kong, Macao, and Taiwan. Benefits have been perceived to be unequally yoked though due to constraints in liberalisation within ASEAN which prohibited some carriers from being able to serve China from any member state within ASEAN, whereas Chinese carriers had the capacity to serve any ASEAN port from any point within China (Lenoir and Laplace, 2016). Thus, the market access for the Chinese carriers was less restrained, providing greater competitive benefit. Lenoir and Laplace (2016) highlighted that there was market dominance by Chinese carriers on the routes between ASEAN and China with Chinese carriers each operating a higher number of routes in comparison with their ASEAN counterparts. According to the authors, the ASEAN carriers reportedly operated 250 routes in 2015 relative to the 370 operated by the Chinese side. Consequently, research by Liu and Oum. (2018) advocated a policy shift which facilities a "made in china LCC" policy by encouraging LCC start-ups in over 80 major cities. Liu and Oum (2018) have gone further to contend that increasing liberalisation out of China magnifies the potential for Chinese carriers to emerge as global mega carriers.

More pointedly though, this first iteration of the Air Transport Agreement (ATA), officially regarded as the first protocol created opportunities for new carriers to enter into the market with more favourable benefits to the Chinese side. Eight new Chinese airlines established services subsequent to the establishment of the ATA far exceeding two new ASEAN entrants. With respect to the increasing number of carriers, the seemingly limited investments by the ASEAN side is not necessarily indicative of a lack of interest though since 27 ASEAN carriers in the market exceeded the seven on the Chinese side. Notwithstanding the competitive diversity the ASEAN carriers brought to the market, the additional Chinese entrants further solidified China's dominance on the network.

A second protocol in 2015 extended the market access provisions affording carriers from either side access to fifth freedoms with some capacity constraints. The agreement provided fifth freedoms "up to 14 services from any of 10 points in ASEAN via any point outside ASEAN/ China to any of the 10 named points in China and beyond to any point outside ASEAN and China" and vice versa. The agreement defined the ten points designated as follows: in the case of the ten ASEAN points,

> Bandar Seri, Begawan, Preah Sihanouk, Mataram, Luang Prabang, Kota Kinabalu, any one point in Myanmar except Yangon, one point in the Philippines except Manila, Singapore, Chiang Mai, any one point in Viet Nam except Ha Noi and Ho Minh City.

The 28 points in China were Changchun, Changsha, Chengdu, Chongqing, Dalian, Fuzhou, Guilin, Guiyang, Haikou, Harbin, Hohhot, Kashgar, Kunming, Lanzhou, Lhasa, Sanya, Shenyang, Nanning, Ningbo, Urumqi, Wuhan, Xi'an, Xiamen, Xining, Xishuangbanna, Yanji, Yinchuan, and Zhengzhou. The curtailed provisions in protocol two primarily signified through the limited scope and focus on secondary cities, reflect a somewhat protectionist stance in relation to liberalising fifth freedoms. However, the protocol nonetheless created market potential in the case of China–ASEAN–ASEAN (ASEAN–ASEAN–China) and ASEAN–China–ASEAN routes.

However, fifth freedoms beyond the routes entailed within the ASEAN–China coalition naturally presents a challenge for China's domiciled carriers. As explored in the previous chapters herein, internal sentiments have fostered views of the carriers as not being equally competitive relative to other international carriers. Thus, China was apprehensive about liberalising access from the three major Chinese cities (Beijing, Shanghai, and Guangzhou) which undoubtedly would have been of appeal to ASEAN carriers. China was able to capitalise on weaknesses derived from further necessary liberalisation within ASEAN itself that would allow carriers to establish services from other member states. While Chinese carriers had the benefit of a wider breadth of origin points, unresolved limitations in ASEAN constrained origins to their domiciled states. From the Chinese side, the extent to which Chinese carriers would gain any consequential benefits from access to broader fifth freedom through ASEAN states would be the primary consideration. Ultimately, the eventuality of these compounded factors was the less liberal adoption of the fifth freedom.

The ASEAN agreement establishes a useful benchmark for the China–Japan–Korea FTA and the resulting air transport implications among these members. However, with slow progress on its implementation and some contentious issues among the three, how air transport factors into the discussion is yet to be fully dissected. Almost six years after its establishment, the trio are still examining how their self-interests can be cohesively aligned. Statements attributed to officials from Korea's Ministry of Trade, Industry, and Energy assert:

> it is not easy to reach a consensus as the three have different sensitive items. China is sensitive to Japan in the manufacturing segment, Japan is to Chinese agricultural products and we (Korea) are sensitive to Japanese manufacturing and Chinese agricultural products.
>
> *(koreaherald.com)*

The FTA once resolved promises to give rise to the world's third largest trading bloc. Viewed in these terms, in comparison with the China–ASEAN FTA and the resulting disposition of China and its air transport dominance, it remains to be explored how the China–Japan–Korea FTA further implicates air transport in the region.

Impact on China's international air transport policymaking

China's involvement in regional political and economic collaboration only started in the late 1990s. Ever since then, China has prioritised bilateral free trade negotiations as one of its top agendas in its overall international strategic considerations. It has demonstrated a magnificent interest in joining regional organisations with an overwhelming enthusiasm in pursuing free trade with its trade partners at both regional and global levels (Hoadley and Yang, 2008). Statistics showed that by the mid-2000s, China was talking with 27 countries and regions regarding the establishment of nine FTAs or Closer Economic Partnership Agreements, covering one-fifth of its total trade (Hoadley and Yang, 2008). By 2018, it had signed FTAs with ASEAN member states and 15 other nations including Chile, Pakistan, Singapore, and New Zealand. An additional 24 FTAs are still under consideration, which included countries such as Canada, Bangladesh, and Mongolia. Proposals were also put forward to Japan and South Korea (now through the China–Japan–South Korea FTA), India and other nations in the Shanghai Cooperation Organisation (SCO).[2]

Historically, China has had close air links with ASEAN member states, with Thailand, Laos, and Malaysia being one of the first to establish bilateral air transport relations. Before the late 1990s, airlines in both China and the Southeast Asian countries had not developed a sound hub-and-spoke system, with point-to-point bilateral markets being the focus of their operations. Governments regarded each country-pair market as an independent operation, hence restricting relaxed arrangements including fifth freedom traffic rights. Carriers from the USA, UK, and the Netherlands were more interested in their expansion across the Atlantic, leaving the Asian market as a second or third priority. In addition, the financial crisis in the region in the late 1990s had affected the air transport market, resulting in low traffic demand with airlines suffering big losses.

Surviving the crisis, the air transport industry in Asia has maintained a remarkable growth rate. In the 2015–2035 IATA passenger forecast, Asia-Pacific was expected to account for the largest number of air passengers globally growing at 4.7% per annum to a total of 3.1 billion passengers. It would lead European and US markets with 1.5 billion and 1.3 billion passengers, respectively. Regional economic integration requires close collaboration with each other, with air transport being a high demand to facilitate the harmonised process. Since 2002, China has held numerous discussions with ASEAN member states including Indonesia, Cambodia, Laos, Singapore, Malaysia, Thailand, and the Philippines, resulting in significant expansion of air services arrangements in terms of the number of designated carriers, capacity, destinations, codeshare arrangements, and fifth freedom traffic rights. In 2004, China and Thailand agreed an open skies accord which allowed unlimited operation of services between the two countries. In the same year, a total of 23 airlines, including LLCs such as Tiger Airways, Air Asia X, and Jetstar Asia, were operating on 104 routes with 481 weekly frequencies between China and ASEAN countries, covering most of the major cities in the region (CAAC, 2004a).

Although the air transport relationship between China and the ASEAN members does not have a direct impact on China's consideration with respect to its policies towards the China–USA, UK, and Netherlands markets, respectively, China's changed mindset on liberalisation with a proactive approach towards the China–ASEAN free trade zone and its relaxed arrangements with ASEAN member states has served several objectives for the country:

- The participation in open skies arrangements has enabled China to accelerate the marketisation of its economy and industry at large. CAAC is able to gain experience in bargaining with its trade partners for optimised benefits for its industry and society as a whole;
- The phased-in arrangements for establishing an open skies zone has allowed its airlines to develop new strategies to adapt to the new environment and to build up its strengths for further expansion;
- The promotion of a regional free trade mechanism has exposed Chinese business to new competition as a result of the rationalisation of market forces. Such competition has helped Chinese airlines to improve their efficiency and productivity, hence gaining experience in competing in the more dynamic world;

- The arrangements have enabled Chinese airlines to expand their networks into peripheral countries to strengthen their competitiveness in the region;
- The arrangements have enabled Chinese airlines to strengthen their regional hubs for further expansion. Airports in Guangzhou, Shenzhen, Shanghai, Chengdu, and Chongqing have had more international flights to serve the region, providing more point-to-point services as well as connecting services;
- The implementation of the liberalised arrangements and the performance of Chinese carriers would enable the government to assess whether the strategies and arrangements have brought about the benefits to the industry and society it wanted to, hence enabling it to adjust its strategies to suit its needs.

Geopolitics and the ASEAN aeropolitical environment

Open skies between the USA and China's peripheral states

In Asia, the US open skies initiative began in the summer of 1996, with Macau SAR being the first to conclude such an agreement, followed by Singapore, Brunei, the Philippines, Taiwan (China), South Korea, Malaysia, Laos, Vietnam, India, and Japan (www.dos.gov). One of the features worthy of note is that all the arrangements allowed seventh freedom traffic rights for cargo operation with limits remaining on fifth freedom traffic rights for passenger traffic. In May 2001, the USA and four other Pacific Rim nations (Singapore, Brunei, New Zealand, and Chile) signed an accord known as the "Multilateral Agreement on the Liberalisation of International Air Transport," with a new element being the full liberalisation of cargo operations between the five countries (Zhang and Zhang, 2002). Looking around at its neighbours, China has been surrounded by open skies agreements and is singled out as the only country in the region that has not concluded any open skies with any other country of the world.

USA–Japan open skies

Japan signed an ASA with the USA in 1952, which enabled the USA to designate three carriers, namely Pan Am (later replaced by United), Northwest (now part of Delta), and Flying Tiger (later acquired by FedEx) to operate to Tokyo, Osaka, and Naha (after Okinawa was returned to Japan) and use these airports to operate to the rest of Asia including China with full fifth freedom traffic rights. In contrast, Japan was only able to designate one carrier, namely Japan Airlines (JAL), to operate to seven points in the USA, e.g. Anchorage, Honolulu, Los Angeles, New York, and San Francisco, as well as Saipan and Guam, with fifth freedom traffic rights being allowed for only one beyond point (Yamauchi, 1997).

Soon afterwards, the Japanese government complained about the inequality of the traffic rights and the advantage enjoyed by US carriers, arguing that such an asymmetric arrangement had hampered fair competition between the two countries, with the US side enjoying more flexibility in terms of designation, traffic right entitlement, and capacity provisions, which put the Japanese side in a disadvantaged position. The following years saw several discussions between the two countries (in 1976, 1985 and 1989) but without much amendment to the 1952 accord (Endo, 2007). In the discussions held in 1997, the US pressured the Japanese government for a liberalised deal that would scrap all restrictions allowing any airline to fly wherever it wanted between the two countries and beyond to other destinations in the fast-growing Asian market (Economist, 1997a). The rationale was that the USA wanted to capture as much as it could of the lucrative Asian market, whose share of air traffic was predicted by IATA to rise from a quarter in 1993 to over half by 2000. Japan was particularly attractive for the USA in that the total passenger traffic between Japan and the USA increased from 101,505 in 1990 to 160,583 in 1999 (Endo, 2007) (Table 7.1), which was one of the biggest and most profitable businesses for US carriers. In addition, the USA saw the potential of intra-Asia traffic where Japanese travel had increased by more than 10 percent until 1997 (Economist, 1997b). The USA estimated that the prospective deal would mean an additional $4 billion in revenues for US airlines over the following four years and save US passengers $1.2 billion as a result of the greater competition and increase in frequencies (Lord and Noah, 1998).

However, the liberalised proposal was rejected by the Japanese government, which was concerned that the Japanese carriers would not receive the same treatment as the US ones do in the Japanese/Asian market, nor were they able to take advantage of the relaxed arrangements to optimise their operations in the US/North American market. The hard negotiation ended up with a compromise that allowed US carriers to increase frequencies to Japan and Japanese airlines to have more gateways into the USA. Although more carriers, such as American, Delta, Continental and UPS from the USA and All Nippon Airways (ANA), Nippon Cargo, and Japan Airlines System (JAS) from Japan, were able to enter

TABLE 7.1 Total passenger traffic between Japan and the USA from 1990 to 1999

	Traffic from Japan to the USA		*Traffic from the USA to Japan*	
	1990	1999	1990	1999
1	8,231	14,412	22,070	31,834
2	12,411	18,380	16,457	27,090
3	8,103	12,192	17,230	31,632
4	2,300	3,960	14,703	21,083
Total	31,045	48,944	70,460	111,639

Source: Adapted from Endo (2007).

(*Remarks*: No.1 refers to both US and Japanese citizens taking US and Japanese carriers. No.2 is US and Japanese citizens taking foreign carriers. No.3 is the traffic of foreign citizens taking foreign carriers, while No. 4 is the traffic of foreign citizens taking US and Japanese carriers).

the market, such an arrangement still resulted in the disappointment of other US carriers such as Northwest, which carried the most traffic between the countries but failed to implement its strategies to leverage Japan as a stop to pick up traffic to destinations further and beyond the region (Economist, 1997a).

US carriers were capable of expanding their regional networks to cover more points in the rest of the Asian countries by optimising their traffic rights entitlement in Tokyo and Osaka. Statistics showed that more than 2.3 million passengers travelled on intra-regional routes, including Asia and Oceania via Tokyo as fifth freedom traffic, accounting for 24 percent of the total intra-regional traffic handled at Narita airport (Oum and Lee, 2002). US carriers were able to capture 20 percent of the lucrative Tokyo–Hong Kong market, with their total market share in Asia increasing from 58 percent in 1986 to over 70 percent in the 1990s.

Another decade of hard negotiation eventually brought about the USA–Japan open skies deal in October 2010, and the long-standing goal of the USA was eventually achieved (Centre for Asia Pacific Aviation, 2009). The deal erased limits on the number of flights, designation, and capacity between the two countries (Cooper and Matsuda, 2010) and allowed unlimited opportunities for cooperative marketing arrangements, such as codeshare, and enabled the carriers of both sides to apply for anti-trust immunity. The deal would allow carriers to work "more closely in setting fares, arranging flight schedules and routes, and coordinating on cost savings through increasing codeshare flights and other operational methods", as elaborated by Tokyo-based Japan Aviation Management Research (Taipei Times, 2010). In addition, US carriers were guaranteed access to Tokyo's Haneda Airport for the first time in the 30-plus years that it had been closed to US carriers. As a consequence, Delta, American, and Hawaiian Airlines were able to launch their long-awaited Japan-bound passenger services, with UPS and Polar Air Cargo expanding their freight services. According to the analysis of the Centre for Asia Pacific Aviation, data available in March 2011 showed that JAL operated a total of 792 international flights per week with 199,078 seats, while ANA flew 541 weekly international frequencies with 129,281 seats. However, the two only accounted for 22.3 percent of the total international capacity to/ from Japan. In contrast, Delta, after taking over Northwest in 2008, which had a hub in Tokyo for its Asia-Pacific operations, operated 508 weekly flights with 135,944 seats, making itself Japan's second biggest international carrier. Together with others, US carriers managed to take 14 percent market share. Of the eight Chinese carriers, China Eastern offered around 52,480 seats per week, ranking the seventh in the league.

Specifically, in the context of the Japan–USA market, Japanese carriers trailed their US counterparts in the market. In the week commencing 18 September 2011, Centre for Asia Pacific Aviation data revealed that Delta led the top four carriers with 35,865 seats per week with 114 departures on Japan–USA routes followed by Japan airlines' 21,575 seats (70 departures), ANA's 16,666 (56 departures), and United Airlines' 15,512 (56 departures). In the week commencing 17 September 2018, although these remained the

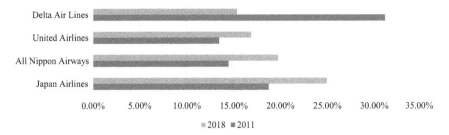

FIGURE 7.1   Market share of the top four carriers between Japan and the USA measured by seat capacity, Sept. 2011 vs. Sept. 2018.

Source: centreforaviation.com.

top four airlines in the market, the two Japanese carriers had overtaken their
American counterparts as market leaders of the country pair (see Figure 7.1).
In 2011, of the 114,918 available seats and 382 departing frequencies, only
33.3% (38,241) of the weekly seat capacity was attributed to the Japanese carriers and 60% (67,796) to the five American carriers plying the routes (American
Airlines, Continental, Delta, Hawaiian, and United). In 2018, however, the
market share of the two Japanese carriers had grown significantly accounting
for 44.8% (54,148) of weekly seats. This increase closed the gap in market
share relative to their US counterparts who had now accounted for only 48.8%
(58,899) of weekly seat capacity.

USA–South Korea "open skies"

The USA and South Korea concluded their ASA in April 1957. The MOU signed
in September 1978 removed the restrictions on designation and pricing, allowing
multiple airlines to operate both scheduled and charter services between the two
countries with the freedom of setting fares based on commercial considerations
with minimum intervention from the governments. The MOU also agreed that
Korean carriers could operate to the USA via points in Japan and that US carriers could serve the Korean market and beyond points via intermediate points.
Following the adoption of 1990s "Guidelines of Supervision and Development
of the Multiple National Carriers" which enabled the accommodation of the
interests of both KAL and Asiana in their international expansion under the constraints of bilateralism, the Korean government agreed to revise the Korean–US
ASA once again in June 1991, which eliminated the constraints on route designation and pricing control, thus creating a pro-competitive bilateral environment
between the two countries.

    South Korea essentially granted unlimited fifth freedom traffic rights to
US carriers. In exchange, the USA allowed Korean carriers to increase their
US entry points, access to cargo terminals, and ground services (Oum and

Lee, 2002). The open skies agreement was eventually concluded in June 1998, permitting unrestricted air services by the airlines of both countries to be operated between and beyond the other's territory, eliminating all the restrictions with respect to flight frequencies, routing, destinations, and pricing (Oum and Lee, 2002).

Having aggressively promoted air transport liberalisation, South Korea hopes that its efforts will help to strengthen its regional hubbing status and enhance the competitiveness of its industry. With its successful conclusion of open arrangements with Japan, another key air transport market in the Asia-Pacific, and more relaxed arrangements with China with respect to the operations in Shandong Province where South Korea has the majority of its investments, South Korea was able to take advantage of its gains through such arrangements and attract traffic between China and the US market, hence generating significant pressure on the Chinese side which is keen to promote its airports in Beijing and Shanghai and on those megacarriers that are also serving the China–Korea, China–Japan, and China–USA markets.

USA–Hong Kong SAR open skies

As a special administrative region (SAR) of China, after 1997, Hong Kong was authorised to negotiate its own bilateral agreements with any third country, although this would have to be agreed to by the central government. This principle applies to air transport arrangements. The 2002 accord with the USA following three years of discussions allowed US carriers to have fifth freedom traffic rights at Hong Kong, enabling them to operate 58 frequencies per week to Japan, South Korea, and other Southeast Asian countries and regions, including the Philippines, Thailand, and Taiwan, as well as to carry freight to the rest of the world, though excluding Australia and Japan. Notwithstanding the strong opposition of Cathay Pacific, which argued that the relaxation only gave favourable consideration to US airlines without considering Cathay's request to fly to South America and Europe via the USA, the Hong Kong SAR finalised the agreement with the central government's acknowledgement. This was an indication that China was getting ready for the move towards embracing relaxed air transport agreements that were not necessarily favouring its own industry.

Impact on China's international air transport policymaking

Northeast Asian integration lags far behind that in Southeast Asia, although there is no shortage of interest in forging closer regional ties (Ruan, 2006). Among the three countries, South Korea takes a more progressive approach in terms of trade liberalisation despite placing less emphasis on economic cooperation (Aggarwal and Lee, 2011). It is one of the few major economies to enter into a FTA with the USA to eliminate 99.7 percent of its existing tariffs. Japan, on the other hand, tends to conclude asymmetric agreements by making fewer concessions vis-à-vis its FTA partners. As it has a clear preference for enhancing economic partnership that goes well beyond trade liberalisation, Japan seeks to incorporate provisions about intellectual property rights, financial services, and investment into an all-embracing package (Aggarwal and Lee 2011). In contrast, China prefers a narrow and simple FTA in terms of scope and coverage so as to reflect its underdeveloped regulatory framework in certain industries. It restricts trade liberalisation primarily to trade in goods by excluding provisions or delaying liberalisation in services (Aggarwal and Lee, 2011).

The call for an open Northeast Asia area not only came from countries such as South Korea, but also from academics such as Oum and Lee (2002), who suggested creating a "Regional Trade and Transport Facilitation Committee," namely a China–Korea–Japan trading block, which could be used as a platform to negotiate compromised agreements to cover the trade and transport fields, since the senior officials and experts of the committee would be in a better position to balance the trade-offs with each other. Tactical measures have been suggested to deal with countries like China which tend to be protective of its industry and action plans have been proposed, including provision of incentive systems, involving consumers, and making use of other ad hoc measures such as appointing special envoys to represent each government to negotiate the agreements (Oum and Lee, 2002).

The Japan–China–South Korea Cooperation was initially launched in 1999 with the primary objective of facilitating the coordination and realisation of various cooperative projects to promote free and fair trade for mutual economic benefits. The Joint Declaration on the Promotion of Tripartite Cooperation between the Republic of Korea, Japan, and China in October 2003 helped to speed up the collaboration. Over the years, 17 ministerial meetings as well as over 50 exchange programs and dialogues have been held, which have significantly enhanced the cooperation between the three parties in various fields including transport and logistics. In September 2006, the first Trilateral Ministerial Conference on Maritime Transport and Logistics was held, with the release of a joint statement of the Japan–China–Korea Ministerial Conference on Maritime Transport and Logistics and two attachments (the framework of the China–Japan–Korea Ministerial Conference on Maritime Transport and Logistics and the plan of action for the China–Japan–Korea Ministerial Conference on Maritime Transport and Logistics) (Ministry of Foreign Affairs of Japan, 2007). The aim was to promote a secure, efficient, cost-effective, and seamless logistics system in Northeast Asia.

Air transport has been excluded in this mechanism due to its complexity in negotiating bilateral agreements under the Chicago regime. However, the ASEAN plus three mechanism as well as the bilateral senior level forum and dialogue have lent momentum to the three parties, which have been actively pursuing the possibility of establishing a more relaxed regulatory and operational environment within the region. The interdependent economies and the trade with each other have laid good foundations for further negotiation for a liberalised environment, which is anticipated to be achieved under the regime of a

Northeast Asia Free Trade Area (NAFTA), so as to allow member states to enjoy more favourable trade and investment conditions.

Despite lacking a regional forum to discuss the air transport arrangements trilaterally, the agreements are very relaxed, allowing multiple designations, codeshare arrangements, and flexible pricing arrangements. As a consequence, air traffic between the three countries grew remarkably in spite of the external uncertainties such as the Asian financial crisis in 1997, SARS in 2003, and the 2008 Global Financial Crisis. Major carriers such as JAL, Korean Air, and China Eastern invested more than two-thirds of their international capacity into the region to capture the market demand (Figure 7.2).

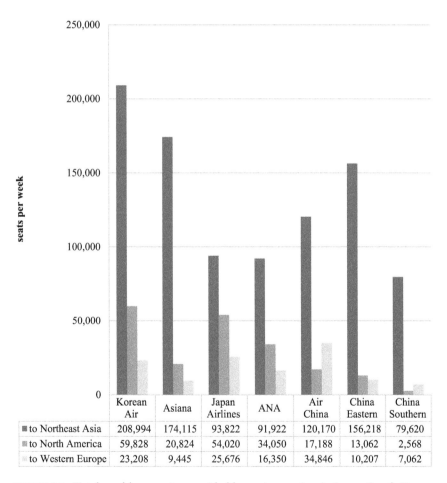

	Korean Air	Asiana	Japan Airlines	ANA	Air China	China Eastern	China Southern
■ to Northeast Asia	208,994	174,115	93,822	91,922	120,170	156,218	79,620
■ to North America	59,828	20,824	54,020	34,050	17,188	13,062	2,568
▫ to Western Europe	23,208	9,445	25,676	16,350	34,846	10,207	7,062

FIGURE 7.2 Total weekly capacity provided by major carriers in Japan, South Korea, and China to Northeast Asia, North America, and Western Europe (March 2011).

Source: centerforaviation.com.

Such liberalised arrangements, on the one hand, satisfied the demand for business and leisure travel as well as bilateral trade, while on the other hand, gave birth to opportunities for airlines in Japan and South Korea to grab traffic between China and the USA. Historically, Japan has always been involved in the Sino-US market. Carriers of both China and the USA had been allowed to use airports in Japan such as Tokyo Narita as intermediate points between the two countries although without commercial traffic rights until 1995, when the Arctic route was opened for direct operations. Allowing US carriers to establish a regional hub in Japan has enabled them to have a wider access to the Asia-Pacific market including China. For South Korea, although the country has not been included in the route schedules between China and the USA, the relaxed arrangements between South Korea and China, and between South Korea and the USA have enabled Korean airlines to take advantage of such arrangements. They were able to create more traffic between China and the USA via their home hubs, as John Jackson, Korean Air's North America marketing director noted that

> US carriers cannot expand their frequencies like this (near doubling of Korean Air flights) due to the limited traffic rights. We have open skies with China and an enormous North American network. This gives us an edge and supports our goal of becoming the preferred airline to China.
>
> *(Steinberg, 2007)*

Permitting carriers of Japan and South Korea to fly to a handful of secondary airports in China enabled them to optimise their traffic rights to feed their hub operations such as at Tokyo and Incheon, contributing a proportional amount of traffic between China and the USA. The convenient transit connection provided to passengers destined to both China and the USA and beyond points, thus, diluted the attractiveness of Chinese operations in Chinese airports such as Beijing, Shanghai, and Guangzhou.

Wolf (2001) argued that regulation of an air transport market generates spill-over effects to other markets that are interconnected by route networks. Traffic diversion benefits countries that have already liberalised their air transport links not only on routes that are substitutes for the regulated one within the same market, but also in other markets. Consequently, a country that still executes regulation in the presence of liberalised interconnected markets runs the risk that its air transport links degenerate to merely spokes of hub operations which are located in other countries (Wolf, 2001). The impact of this kind of traffic diversion is not only on airlines and airports but also on governments, as airlines are struggling to maintain their levels of traffic to achieve economies of scale, airports are threatened to lose their hub status, and governments are concerned about the competitiveness of their aviation industries compared with their counterparts.

The case of the China–Japan–South Korea–USA relationship in the Sino-US air transport market reflects the above proposition. The open skies arrangements between the US and Chinese peripheral economies and between Japan and South Korea, which was concluded in August 2007 that lifted restrictions on frequencies, capacity, and destinations between the two countries, though with the exception of the congested Tokyo airports, have exerted significant pressure on the Chinese government, airlines, and airports. For the Chinese government, it is concerned how its major carriers will compete with their rivals in the region. It faces pressure from such major airports as Beijing, Shanghai, and Guangzhou demanding more liberalised arrangements. It understands that relaxing the economic restrictions is inevitable, but the key issue remains how quickly the market should be relaxed. In the case of the bilateral discussions in the 1980s and 1990s when the concept of "open skies" had only just started to be picked up in Asia, China did not see the necessity for a liberalised environment for its businesses, being convinced that intervention was imperative to guide the country through adverse economic cycles, such as the financial turmoil experienced in Southeast Asia in 1997.

While in the 21st century, open skies has become a main theme in every bilateral negotiation, the consideration is not whether to acknowledge it or not, but to what extent it should be accepted. For airlines, they are disadvantaged in developing their hubs at their home bases to extend their networks, losing both direct and indirect traffic due to the connecting passenger traffic at third countries' hub airports. Statistics showed that around 16 percent of USA–China passenger traffic in 2006 was lost to third–country carriers (Steinberg, 2007). For airports, especially major airports such as Beijing, Shanghai, and Guangzhou, they experienced less demand from connecting passengers as a result of their insufficient network coverage. Their strategic objective of being the key hubs in the Asia-Pacific region is at stake.

Before the Sino-US negotiations in 2004 and 2007, Chinese airports presented a strong business case pointing out that they had experienced a moderate growth in terms of direct traffic between China and the USA, with a proportion of their direct traffic taken away by their Korean/Japanese counterparts. They faced the risk of losing the battle to be major hubs in Asia to Incheon, as a result of being unable to capture the Sino-US market. The message was acknowledged, which noted that one of the objectives in relaxing the Sino-US market was to counter the competition from Korean and Japanese carriers, which have taken advantage of the fifth and sixth traffic rights enabled by the open skies arrangements between themselves and the USA.

To summarise, liberalised arrangements between China's peripheral states and the USA, between themselves, and between China and its neighbours have had a significant impact on China's international air transport policymaking. Such an impact, though, is more on policies with respect to the specific country-pair markets rather than the overall national liberalisation policy initiatives. This has been reflected in the arguments presented by Chinese interest groups (discussed in the following chapters) before each round of discussions, especially between China and the USA.

Notes

1 ASEAN Plus Three Summit is an informal summit of the ten ASEAN member states, China, Japan, and the Republic of Korea (South Korea).
2 Shanghai Cooperation Organisation was an intergovernmental organisation founded in Shanghai, China in June 2001 with member states, including China, Kazakhstan, Kyrgyzstan, Tajikistan, Uzbekistan, and the Russian Federation. The primary goal is to manage the potential Sino-Russian tensions or competition with overt activities being directed first at transnational threats as well as cooperation in areas such as economy and trade, science and technology, culture, education, energy, transportation, and environmental protection (Bailes et al., 2007, and www.fmprc.gov.cn)

References

Abeyratne R., (2003), The worldwide air transport conference of ICAO and its regulatory and economic impact. *Air and Space Law*, XXVIII/4–5, September 2003, 218.

Aggarwal V. K., and Lee S., (2011), The domestic political economy of preferential trade agreements in the Asia-Pacific, in: Aggarwal V. K., (edited), *The political economy of the Asia Pacific, The role of ideas, interests, and domestic institutions*, (pp. 29–48), Springer, New York and London.

ASEAN, (2003), Yearly archives: 2003, available at https://asean.org/2003/

Bailes A. J. K., Dunay P., Guang P., and Troitskiy M., (2007), The Shanghai Cooperation Organisation, Stockholm International Peace Research Institute (SIPRI) Policy Paper No. 17.

Basu Das, S., (2018), Seamless ASEAN sky: Policymakers need to look beyond obstacles, available at https://www.iseas.edu.sg/media/commentaries/seamless-asean-sky-policymakers-need-to-look-beyond-obstacles-by-sanchita-basu-das-2/

CAAC, (2004a), China Civil Aviation Annual Policy Statement for 2003.

CAAC, (2004b), *Statistical data on civil aviation of China, 2004*, Civil Aviation Press of China, Beijing.

CAAC, (2005), China Civil Aviation Annual Policy Statement for 2004.

CAAC, (2006), China Civil Aviation Annual Policy Statement for 2005.

CAAC, (2007a), CAAC's Policy Proposal with respect to the implementation of the Party's Principles of the Sixth Plenary Session of Sixteenth Congress, July 2007, available at www.caac.gov.cn/B7/200707/t20070719_6569.html

CAAC, (2007b), China Civil Aviation Annual Policy Statement for 2006.

CAAC, (2008), China Civil Aviation Annual Policy Statement for 2007.

Centre for Asia Pacific Aviation, (2009), LCCs in Japan, Korea and China: massive growth potential to be realised in the next decade, 27th November 2009, available at www.centreforaviation.com

Chayes A., and Chayes A. H., (1995), *The new sovereignty, compliance with international regulatory agreements*, Harvard University Press, Cambridge, MA and London.

Ching C., and Ching H. Y., (2003), *Handbook on China's WTO accession and its impacts*, World Scientific Publishing Co Pte. Ltd., Singapore.

Cooper C., and Matsuda K., (2010), U.S., Japan sign open skies air travel agreement, 25 October 2010, available at www.businessweek.com/news/2010-10-25/u-s-japan-sign-open-skies-air-travel-agreement.html

Economist, (1997a), Half-open skies? *09/06/97*, 344(8033), 76–77, 2.

Economist, (1997b), An acrimonious dispute between America and Japan over international airlines looks set to get worse, *02/08/97*, 342(8003), 65–67p.3.

Endo N., (2007), International trade in air transport services: penetration of foreign airlines into Japan under the bilateral aviation policies of the US and Japan. *Journal of Air Transport Management*, 13, 285–292.
www.springerlink.com/content/r3jk257016v23911/fulltext.pdf

Hoadley S., and Yang J., (2008), China's free trade negotiations: economics, security, and diplomacy, in: Katada S. N., and Solis M., (2008), (editors), *Cross regional trade agreements, understanding permeated regionalism in East Asia*, (pp. 123–146), Springer-Verlag, Berlin.

ICAO, (1944), Convention on International Civil Aviation, Doc7300/9, available at http://www.icao.int/icaonet/arch/doc/7300/7300_9ed.pdf

ICAO, (2003), Declaration of global principles for the liberalisation of international air transport, available at www.icao.int/icao/en/atb/epm/Ecp/Declaration.pdf

ICAO, (2013), ICAO Working Paper: Expansion of Market Access for International Air Transport in a Proactive, Progressive, Orderly and Safeguarded Manner, available at www.icao.int/Meetings/atconf6/Documents/WorkingPapers/ATConf.6.WP.97.2.1.en.pdf

Kapp R. A., (2002), WTO: towards "year two", letter from the President of the US-China Business Council, *The China Business Review*, 29(6), November–December 2002, available at www.chinabusinessreview.com/public/0211/letter.html.

Kent A., (1999), *China, the United Nations, and human rights, The limits of compliance*, University of Pennsylvania Press, Philadelphia.

Kent A., (2007), *Beyond compliance, China, International organizations, and global security*, Stanford University Press, Stanford, CA.

Kim S. S., (1994), China's international organisational behaviour, in: Robinson T. W., and Shambaugh D. L., (editors), *Chinese foreign policy: theory and practice*, (pp. 401–434), Oxford University Press, Oxford.

Ku S. C. Y., (2006), China's changing political economy with Malaysia: a regional perspective, in: Yeoh E. K., and Chung H. K., (editors), *China and Malaysia in a globalizing world, Bilateral relations, regional imperatives and domestic challenges*, (p. 32), Institute of China Studies, University of Malaysia, Kuala Lumpur.

Lampton D. M., (2001), *Same bed different dreams, managing US-China relations 1989–2000*, University of California Press Ltd., Berkeley, Los Angeles and London.

Lenoir, M., and Laplace, I., (2016), Liberalization in Southeast Asia: who is capturing the markets?. WCTR 2016, *World Conference on Transport Research*, Jul 2016, Shanghai, China, pp. 1–13.

Liu, S., and Oum, T. H., (2018). Prospects for air policy liberalization in China as a result of China-ASEAN open skies: Changing role of Chinese mega carriers in global scene and anticipated Low Cost Carrier competition, *Transport Policy*, 72(C), 1–9.

Lord M., and Noah T., (1998), Prying open Japanese skies, U.S. News & World Report, *02/16/98*, 124(6), 50, 1/4,

Ministry of Foreign Affairs of Japan, (2007), 2005–2006 Progress Report of the Trilateral Cooperation among the People's Republic of China, Japan, and the Republic of Korea, available at www.mofa.go.jp/region/asia-paci/pmv0701/report070112.html

Oum T. H., and Lee Y. H., (2002), The Northeast Asian air transport network: is there a possibility of creating open skies in the region? *Journal of Air Transport Management*, 8, (2002), 325–337.

Ruan Z., (2006), Commentary: China's role in a Northeast Asian community. *Asian Perspective*, 30(3), 2006, 149–157.

Sheng L., (2003), China-ASEAN free trade area: origin, developments and strategic motivations. ISEAS Working Paper: International politics & security Issues Series No. 1, 2003.

Steinberg A. B., (2007), The importance of aviation liberalization, Speech delivered at the US-China Business Council Aviation Forum, January, 2007, Beijing, China.

Taipei Times, (2010), Japan, US seek liberalized aviation with 'open skies', Taipei Times, available at www.taipeitimes.com/News/biz/archives/2010/10/26/2003486901/1

Thomson G., (2003), China's accession to the WTO: improving market access and Australia's role and interests, in: Cass D. Z., Williams B. G., and Barker G., (editors), *China and the world trading system, entering the new millennium*, (pp. 68–82), Cambridge University Press, Cambridge.

United Nations Economic and Social Commission for Asia and the Pacific (UNESCAP), (2010), Striving together, ASEAN & the UN, available at www.unescap.org/stat/statpub/Sriving-Together-ASEAN-UN2010.pdf

Wang Y., (2011), Interview conducted via email on 11 February 2011.

Williams A., (2009), *Contemporary issues shaping China's civil aviation policy, balancing international with domestic priorities*, Ashgate Publishing Ltd, Farnham.

Wolf H., (2001), Network effects of bilaterals: implications for the German air transport policy. *Journal of Air Transport Management*, 7, 63–74.

Wu Z., (2007), Progress and challenge in the process of liberalization of international air transport in China, speech delivered at the China Civil Aviation Development Forum on 10th May 2007, Beijing, China.

Yamauchi H., (1997), Air Transport Policy in Japan: limited competition under regulation, in: Findley C., Chia L., and Singh K., (editors), *Asia Pacific Air Transport, Challenges and Policy Reform*, Institute of Southeast Asian Studies, Singapore.

Yeaman D., (2003), The impact of China's WTO accession upon regulation of the distribution and logistics industries in China, in: Cass D. Z., Williams B. G., and Barker G., (editors), *China and the world trading system, entering the new millennium*, (pp. 19–30), Cambridge University Press, Cambridge.

Yean T., (2010), ASEAN open skies and the implications for airport development strategy in Malaysia, in: Brooks D. H., and Stone S. F., (editors), *Trade facilitation and regional cooperation in Asia*, Edward Elgar Publishing Limited, Cheltenham.pp. 23–57

Zhang A., and Zhang Y., (2002), Issues on liberalisation of air cargo services in international aviation. *Journal of Air Transport Management*, 8, 275–287.

8

APPLYING THE MICRO-MACRO LINKAGE APPROACH (PART II)

Analysis of factors at the bilateral level

Factors at the bilateral level and their impact on policymaking

Of the markets assessed, air service relations were established with Japan in 1974 followed by the UK in 1979, the Netherlands in 1979, the USA in 1980, and South Korea a decade later in 1990. The momentum gained since the 1980s which saw the establishment of air links between China and the USA, as well as numerous rounds of bilateral negotiations between China and each country examined, respectively, started to progressively eliminate the economic restrictions on airline operations. Fundamental changes did not take place until the past two decades in the 21st century, during which period of time, at least two to three rounds of discussions were conducted between China and its selective counterparts, resulting in substantial relaxation of the constraints imposed on the operations in air transport markets, with remarkable achievements being widely recognised.

Foreign governments are more or less traditional actors in foreign economic policymaking, often exerting direct influence through bilateral talks and negotiations (Feng, 2006). The degree of influence is largely determined by the two countries' positions in the international system, as well as their relative political power and economic dependency (Feng, 2006). One country's political policy towards the other will be reflected in their bilateral trade talks with negotiation outcomes bearing trade-offs with political significance. The following examines the bilateral relations between China and its regional peripheral neighbours Japan and South Korea, followed by other international partners assessed, namely the USA, the Netherlands, and the UK. It analyses the impact of political and economic relations on China's international air transport policymaking, in particular, the country-specific policy.

China–USA political relations

Lampton (2001) quoted a well-known Chinese idiom to describe the relationship between the USA and China as "same bed, different dreams." He further explained that the processes of economic and information globalisation, along with the development of international regimes and multilateral organisations, have landed America and China increasingly near one another in the same global bed, but their respective national institutions, interests, leadership, and popular perceptions, and the very characteristics of the people, ensure that the two nations have substantially different dreams. Such an underlying dynamic has been the theme since the first decades of post-war USA–China relations and will continue to be so well into the 21st century (Lampton, 2001). It poses significant challenges for both countries as to how to manage diplomacy in areas such as political, military, and economic relations.

China's view towards the USA

When the new China was established in 1949, the nation was not recognised by most of the member states in the United Nations, including the USA. Like other world leaders who held a bipolar perception of the international system headed by the USA and the Soviet Union, China regarded the USA as a threat to its security and adopted a "leaning to one side" policy, choosing to form a "Sino-Soviet long-term relationship" so as to deter the threat from the USA. However, the Sino-Soviet relationship broke up in the early 1960s, leaving China to pursue an independent position from both superpowers until 1978, when China and the USA reproached their diplomatic ties. Since then, China's US policy has changed from "pro-US" in the 1980s to "multipolarity" in the 1990s and "unipolarity" in the 21st century (He, 2009).

Sino-US political relations and strategic economic dialogue (SED)

The debate within the USA has been endless as to what is the most appropriate China policy, which always gains momentum with each new presidential election (Garrison, 2005). The USA's China policy has evolved over the years with each newly elected president, who chooses strategies and tactics that suit the overall national interest, although engagement with China has been a consistent and patient theme (Shi, 2009).

Khalilzad (1999) explained that US engagement, in principle, seeks to maintain and enhance relations with China as much as possible in the various policy realms. It is a tactic rather than a policy as it refers to the means—increased contact and a dense network of relationships—rather than the objectives, although it helps achieve certain objectives. The view holds that deepening China's economic links with the outside world and increasing its foreign trade and investment are the best guarantees of China's stability and will facilitate China's democratisation, thus it is less likely to come into conflict with the USA. The engagement policy has had a great impact on Sino-US air transport relations, which have been progressing slowly but steadily over the years towards full and complete "open skies."

As a result of the political and legal system, the USA's China policy coordination mechanism had been diffused in large part due to the different focus and attention of different institutions before 2006. For example, the DoS, the Department of Treasury (DoF), and the DoC had all been involved in dealing with China on different economic issues, which had shaped the development of different strands of the USA's China policy by engaging in a series of contentious economic dialogues under their own authority (Garrison, 2007).

To centralise management across the various bureaucracies involved in the complex relationship, the strategic economic dialogue (SED) was jointly launched in 2006 by Bush and his Chinese counterpart Hu Jintao, with an aim of enhancing mutual understanding by engaging top-level officials in direct and regular dialogue to discuss any political, economic, environmental, trade, and cultural issues of mutual concern by better coordinating those well-established bilateral dialogues such as the Joint Economic Commission (JEC) and the Joint Commission on Commerce and Trade (JCCT) (Garrison, 2007). Convened semi-annually and led by the Secretary of the Treasury of the USA and a Vice Premier of China, six rounds of SED dialogues had been held by 2008 with 189 agreements reached. In the fifth round of SED dialogue alone in early 2008, 40 agreements were concluded (Diao and Li, 2008). Being considered as a positive force not only for economic ties but also for movement on many other bilateral issues, it has been heralded by the Chinese side as a constructive means of communicating intersecting interests, bypassing bureaucracies, building relationships, and avoiding megaphone diplomacy (Shi, 2009). Likewise, it was regarded by the USA as an effective way to maintain dialogues between top-level officials to lessen the misperceptions and miscommunications for a very productive bilateral relationship. The SED mechanism has helped keep relations progressing during tough times, and helped both sides to manage tensions that inevitably occurred (Paulson, 2008).

Seeing the benefits of face-to-face dialogue which was able to address the economic differences at a political level, the Obama Administration relaunched the existing mechanism, with the aim of building a positive, cooperative, and comprehensive relationship for the 21st century by providing a senior-level forum for direct communications between the two parties to meld the economic and security aspects of America's China policy into one coordinated joint effort (Rogin, 2010). Held annually in both China and the USA, the dialogue enabled both parties to address all matters of mutual concern, such as climate and energy, which is "an economic issue, a diplomatic issue, a development issue, an energy issue, an environmental issue, an agricultural issue and a national security issue...in its full complexity" (Clinton, 2009). Air transport was first included in the SED in 2007. Although negotiations on bilateral air services between the competent authorities are not necessarily scheduled at the same time as S&ED, both sides work under the regime of S&ED with any conclusions of the agreements resulting from the discussions being timed to fit into the S&ED programme.

In 2016, the newly elected President of the USA Trump espoused a foreign policy position which increasingly and more overtly criticised China–USA trade relations. In part, the president continuously reinforced that China was using practices such as currency manipulation to underpin its trade surplus with the USA. Thus, in April 2017, the S&ED was abolished, and four refashioned bilateral mechanisms were established. The economic dialogue mechanism entitled the USA–China comprehensive economic dialogue (CED) was established with the similar intention of addressing a broad cross section of economic issues between the nations. However, the administration argued the CED would be distinguished by its propensity to push Beijing to far more consequential concessions. After its first meeting in July 2017, the Trump Administration was unimpressed by the results of this inaugural CED meeting charging that China remained unwilling to express a more balanced approach to trade relations with expanded market access. From here, the strained trade relations descended into a trade war which has had significant consequences for trade between the two nations as has been further expressed in Chapter 4.

USA–China Aviation Cooperation Programme (ACP)

Economic aid is regarded as an instrument to foreign policy, with the political relationship being one of the most important determinants of the aid flows (Radelet, 2006). Providing capital and technical assistance is believed to be an effective weapon in the ideological war to help win those uncommitted nations that are also underdeveloped and poor (Frieman, 1994). Greater contributions from developed nations must be linked to greater responsibility from developing nations (Lancaster, 2007). Firmly believing in this philosophy, since the Cold War, the USA has been one of the major aid donators to developing countries, though it was criticised for giving little regard as to whether the aid actually was used to support development (Radelet, 2006). In 2002, the Bush Administration announced an unprecedented amount of aid totalling $5 billion for development to be achieved by 2006, arguing that it was a new compact for global development defined by new accountability for both rich and poor. The elevation of development has since become one of the three priorities of US foreign policy, along with defence and promoting democracy (Lancaster, 2007).

The USA–China Aviation Cooperation Programme (ACP) was created in 2004 with the primary objective being to promote technical, policy and commercial cooperation between the aviation sectors of the two countries. Through the cooperation of CAAC and several US Government agencies including the United States Trade and Development Agency (USTDA), FAA, DoC, the US Embassy in China, as well as its 36 prominent members such as Boeing and General Electric Company (GE), this innovative public–private initiative is designed to link the USA and Chinese governments and aviation industries in a multi-faceted programme so as to create an enabling environment to advance US commercial interests that will benefit US exports of aviation products and services.

Between 2005 and 2011, over $5.5 million had been invested to sponsor a wide range of projects including months-long training for executives, visits to the USA, symposiums and seminars, safety and security projects, as well as airport environmental best practice projects. One of the key efforts is the Executive Management Development Training (EMDT), which provides a four-month cutting-edge training for senior managers coming from CAAC's economic regulation divisions, airlines, and airports (Tymczyszyn and Bai, 2009), with one of the core subject matters covered being the US liberalisation experience and what lessons China could learn. The objective was to expose the trainees as future leaders to the best US aviation practices, procedures, and technologies, thus helping China establish a framework for managing the challenge of its continued growth of the aviation sector. Reimold (2009) exclaimed that

> it is important to remember when it comes to aviation cooperation, it is not just about technologies and procedures, it is about people. Through the EMDT programs, we believed that we have taken an important step towards improving our working relationships with the next generation of aviation leaders.
>
> *(Reimold, 2009)*

Other beneficiaries include China Eastern, which was funded for its staff training; CAAC Air Traffic Management Bureau (ATMB) whose executives are trained with respect to safety and general management; Beijing Airport, which will work together with US professionals on construction of its second airport to ensure an environmental-friendly project; and other Chinese airports, which would be familiarised with US best practices in sustainable airport development by making a trade mission to visit US airports such as Washington, New York, Chicago O'Hare, and Atlanta.

In commenting on the progress and achievements made through the ACP, the USTDA's acting director Leocadia I. Zak stated that

> environmental solutions and technologies are critical to sustainable growth in the aviation sector. Working in cooperation with our Government partners and the US and Chinese civil aviation industries, we are helping our aviation systems grow together in mutually beneficial ways. By sharing experiences and best practices with each other, we can achieve results that will strengthen our respective systems and shape the future of commercial aviation.
>
> *(USTDA, 2009)*

Impacts on China–USA negotiations

Sino-US relations are the most important relationship for both countries. From China's perspective, the USA is not only the world leader politically and economically but also the country's second biggest trade partner. It holds the most advanced technology that China needs for its continuous economic growth. For the USA, China represents an emerging market with vast potential for US business, and a vital player in maintaining the world economic order. It also holds a strategic key to maintain a military balance in East Asia, such as addressing issues in North Korea. However, such an understanding does not mean that both have developed a shared view towards each other and the rest of the world. On the contrary, the two still perceive totally different priorities, dangers and strategies when contemplating how best to deal with each other (Cohen et al., 2003). No theories, precedent, or econometric equation is available for either of them to define its respective national interest and how to ensure that policy outcomes would serve the country's interest best. Conflicting value judgments still need to be reconciled by all those that are involved in the policymaking process (Cohen et al., 2003).

Sino-US political relations have had a great impact on the negotiation results of Sino-US air transport arrangements. The bilateral air transport relationship was formulated immediately after the two countries established diplomatic ties in the early 1980s. For the following ten years, there were few discussions held mainly due to the distant relationship with each other. Political and trade tensions developed since the 1980s could be cited to explain why the two failed to pursue a closer relationship further. The first abrupt relaxation came in 1999, almost 20 years after the bilateral ASA was signed, when China was preparing to be admitted to the WTO, to which the USA held the significant influence over its accession. China publicly announced that it was prepared to offer substantial concessions such as eliminating market entry barriers and called for all industries to take actions in this regard. The next biggest thrust to a liberalised arrangement took place in 2004, three years after China's entry into the WTO. Sino-US political relations had been significantly improved after the 9/11 tragedy, with the two countries declaring to be strategic partners. Regular summits have since been held at very senior levels with an aim of forging a strong link between the decision makers through a continuous and direct dialogue. Although political forums/summits are new phenomena in the political economy that have only come into existence since the 21st century, they have played a key role in strengthening the bilateral relationship by facilitating the communication for an extensive engagement with each other. This was particularly attractive for those countries who wanted to further their relationship with China, as "guanxi", (personal relationship) is regarded as the key to securing business success in the Chinese market.

The Sino-US SED/S&ED was created with such an aim. The USA is fully aware of the importance of personal relationships in Chinese business practice and the SED/S&ED was specially designed to provide a forum for senior officials to meet regularly to discuss any issues that cannot be resolved at junior level. When air transport was brought under the mechanism of the SED, the USA aimed at achieving a breakthrough of what it failed to achieve at industry level, as Paulson (2008) observed. In April 2006, meetings to discuss how to develop air transport services commenced. While the USA wanted to have a full open skies arrangement involving unlimited designation of carriers and unlimited cargo operations, China was convinced that it did not want such a deal at this stage due to its insufficient infrastructure and the unused traffic rights available for Chinese carriers from the 2004 Protocol. The discussions did not progress due to the differences between the two parties. However, things started to get moving again when Paulson brought the case to his Chinese counterpart Madam WU Yi who co-chaired the third round of the SED, which resulted in the successful conclusion of 2007 Protocol. The Protocol eliminated all restrictions on the number of cargo airlines and flights, thereby allowing all-cargo airlines to optimise their global route networks. Agreement was also reached on doubling the permissible market capacity and removing restrictions on Chinese carriers that could enter the market. What is more significant is that both parties agreed that discussions about full open skies arrangements should start no later than March 2010, although even as late as 2018, no such agreement had been reached. Paulson recalled in 2008 that

> lower-level negotiations on such (bilateral air services) an agreement had been stalled for some time because China was focused on developing the competitiveness of its domestic aviation industry and limiting international competition. But through the SED, I brought this issue to the attention of my counterpart, then Vice Premier Wu Yi, and explained how the increased exchanges of people and goods between our two countries would strengthen the relationship. Thanks to the agreement, US passenger flights to and from China will more than double by 2012, and air-cargo companies from both countries will enjoy full liberalisation of the industry, including the lifting of restrictions on the frequency and price of flights, by 2011.
>
> *(Paulson, 2008)*

In terms of the CAP programme, although to date direct influence on China's air transport policy is not observable, the impact on the ideology of aviation officials in China could be long-term and more fundamental. Outside ideas could lead to a shift in the dominant thinking of decision makers and influence their beliefs and behaviours, and ultimately their policy choices, as Feng (2006) claimed. They will also shape their values on society, the culture and even their practice, resulting in the challenging of existing norms. Williams (2009) observed that the aggressive encouragement of the USA seemed to have paid off. CAAC was more inclined to accept the working procedures and business practices of the USA and endeavours to replicate these to fit China's circumstances. The relationship between the Chinese Ministry of Transportation (MoT) and CAAC, which has been an affiliated bureau since 2008, will eventually replicate to some degree the US aviation management system as reflected in the fact that there are, at the current time, a number of ongoing advisory roles being played in China by the US DoT and FAA.

China–EU political relations

The EU's China policy

Over the years, the EU has developed an intricate web of interregional and bilateral contact covering political and commercial relations with almost every country in the world by having seconded over 100 diplomatic missions and offices around the world with a high profile and considerable economic weight (Dinan, 1999). Although the USA has always been of paramount importance for the EU's external relations given its size, economy, and importance in the world order, the EU recently has given priority to emerging economies including China.

Unlike the USA, whose foreign policy is affected by the president of the day, the EU's foreign policy is formulated by the European Commission, the supranational institution claiming to represent the interests of all individual member states. The EU's foreign policy is usually announced through the commission's communications, where the justification of policy direction and implementation strategies is spelled out. Since 1995, five such China policy papers have been published which have served as guidelines for the EU's strategy towards China.

The first EU China policy, "A long-term policy for China-Europe Relations," was announced in July 1995 (Dinan, 1999), as part of the key components of its new strategy towards Asia. By that time, China had become the EU's fourth largest export market and fourth largest supplier. China's further reform and liberalisation was promising in providing more trade and foreign investment opportunities which the EU had been striving to achieve to support its businesses. To this end and to help raise the EU's international profile, the EU recognised the need to redefine its relationship with China, with whom a long-term relationship should be maintained for a "constructive engagement" to appreciate China's unprecedented economic upsurge (Casarini, 2006). Although the policy was very basic, it has laid a good foundation for the commission to drive the relationship with China in a single integrated framework at the community level. China responded actively to the EU's call with its President Jiang Zemin making his historical trip to Western Europe in the same year and calling for the development of a stable, friendly, and long-term cooperative relationship (Zhang and Zhuang, 2005).

The EU's second China policy, "Building a comprehensive partnership with China," was announced in 1998 and was regarded a milestone in EU–China history. Recognising that China's emergence as an increasingly confident world power was of immense historic significance to both Europe and the international community, the commission called for an update of its China policy in response to the challenges posed by the changes in China. Accordingly, a further engagement with China through an upgraded political dialogue in the international community (Men, 2007) should be achieved and EU–China relations should be upgraded to the same level as those of EU–USA, EU–Russia, and EU–Japan (Zhang and Zhuang, 2005). As a result of the political drive, the first bilateral summit meeting was held in London in April 1998 with a joint statement declaring that China and the EU would "establish a long-term, stable and constructive partnership oriented towards the 21st century" (Zhang and Zhuang, 2005). An annual summit meeting mechanism was thus established to facilitate the development of relations.

The partnership was enhanced in September 2003 with a third China policy, "A maturing partnership: shared interests and challenges in EU-China relations," being published to cover the years from 2003 to 2006. Recognising China's growing importance in world politics and economy, the EU realised that an outspoken comprehensive China strategy was required to define the objectives and determine the actions and measures for effective implementation. A comprehensive, independent, and consistent long-term strategy was thus formulated emphasising shared interests not only in bilateral relations but also in global affairs. It stated that "the EU and China have an ever-greater interest to work together as strategic partners to safeguard and promote sustainable development, peace and stability" (Men, 2007). A total indicative budget of €250 million was envisaged for the five-year period covering various projects and activities in different sectors, such as human resource management, environment, and sustainable management (Council of the European Union, 2006).

The EU's fourth China policy, namely, "EU-China: Closer partners, growing responsibilities," was detailed in 2006 in the commission's communication to the council and the European Parliament, in which the commission explicitly proposed that a dynamic relationship with China was needed to allow leverage based on the values of the EU (Council of the European Union, 2006). It declared that the EU should continue to support China's internal political and economic reform process for a strong and stable China which fully respects fundamental rights and freedoms, protects minorities, and guarantees the rule of law. Further, it was expressed that the EU should also reinforce its cooperation with China to ensure its sustainable development, pursue a fair and robust trade policy, and work to strengthen and add balance to bilateral relations. Co-ordination and joint actions should also be increased to improve cooperation with European industry and civil society (Council of the European Union, 2006). A number of key objectives were thus identified, some of which are as follows:

1   Supporting China's transition towards a more open and pluralist society;
2   Promoting sustainable development;

3   Ensuring secure and sustainable energy supplies;
4   Combating climate change and improving the environment;
5   Improving exchanges on employment and social issues;
6   Improving coordination on international development; and
7   Building sustainable economic growth (Council of the European Union, 2006).

For the effective implementation of the policy, a "China Strategy Paper" was published by the Council of Foreign Relations in 2007 to cover the seven-year period from 2007 to 2013. Recognising the "duality" of China, namely a developing country in terms of certain traditional indicators on the one hand but a significant player on the world stage in economic and political terms on the other, the EC's response strategy towards China was adjusted to be targeted at:

1   Providing support for China's reform programme in areas covered by sectoral dialogues;
2   Assisting China in tackling global concerns and challenges over the environment, energy, and climate change; and
3   Supporting China's human resource development (Council on Foreign Relations, 2007).

An indicative funding of €224 million for the period was allocated with the assistance in the areas of intervention being complemented by actions and support to be provided through various thematic programmes and regional budget lines (Council on Foreign Relations, 2007), with a budget totalling over 20 million euros to be allocated to civil aviation.

In July 2016, a new strategy entitled the EU–China 2020 Strategic Agenda for Cooperation was adopted. Recognising the EU's importance as China's largest trading partner, the agreement sought to affirm cooperation between China and EU members in areas of peace and security, prosperity, sustainable development, and people-to-people exchanges. In forging greater ties in peace and security, the agreement would see China and the EU strengthen their cooperation and coordination in advocating for equitable and effective global rules in areas such as international trade and investment, nuclear security, human rights, and maritime security, among others.

To ensure sustained mutual prosperity, key developments in trade and investment were proposed including an EU–China Investment Agreement which in part opens market access and advances economic reforms in China. Greater prosperity would also be achieved through enhancements to the EU–China Industrial Dialogue and Consultation Mechanism with the overall intention of improving information sharing and industry growth. The agreement also recognised the importance of enhancing cooperation in areas of agriculture to provide enhanced food security. Most important in the context of aviation was the declared intent to gain further progress on matters related to transport and infrastructure one of which was to enact a letter of intent signed in 2013 to enhance cooperation in aviation. The third pillar of the policy was built on ensuring sustainable development in areas such as science and technology, space and aerospace, and climate change, while the final pillar was premised on enhancing cultural interaction between the respective citizens by providing greater opportunities for people-to-people exchanges.

The EU seeks to promote regional stability and global prosperity by achieving "smooth and gradual integration of China into the world economy." The commission and member states are trying to achieve greater coordination of EU activities and national politics relating to China, and a higher EU profile in China and throughout Asia (Dinan, 1999). The policies published at different times have helped the EU to refine its policy towards China by identifying objectives and proposing the most appropriate implementation strategies which have safeguarded the China–EU relationship over the past three decades.

China's view towards the EU

Enhancing China–EU relations is an important component of China's foreign policy. The Chinese leaders and scholars believe that the EU holds different views from the USA regarding the future of the international order which is quite close to that of Chinese thinking (Wai, 2005). The EU favours multilateralism in addressing international issues by emphasising common interests, sharing of power and abiding by the restrictive common rules of the games (Wai, 2005). From the Chinese point of view, the EU is considered a major actor/partner in the formulation of a new order for the benefits of China. As a strategy to combat the influence of the US, China first put forward its multipolarity policy in the late 1970s. In 1998 when the first EU–China summit was held, China proposed to use the term "multipolarity" in the final joint declaration, which was refused by the British representative who was concerned about misinterpretation from the US side, which might cause unnecessary misunderstanding, thus damaging the already weak relationship between the EU and USA across the Atlantic (Men, 2007). Though China chose not to replace the term multipolarity, its denotation was interpreted differently by highlighting the elements of multilateralism such as democratisation of international relations and the strengthening of coordination and dialogue (Men, 2007). In October 2003 as a response to the EU's China policy paper, China released its first-ever "EU Policy Paper" outlining plans and measures on bilateral cooperation for the next five years.

Recognising the EU as a major force both politically and economically in the world, the paper stated that the common ground between China and the EU far outweighed the divergent views on certain issues as there existed no fundamental conflict of interest between the two parties (FMPRC, 2003). On the contrary, China and the EU were highly complementary economically in that the EU had a developed economy, advanced technology and strong financial resources, while China boasted a steady economic growth, a huge market and abundant labour force. To strengthen and enhance the China–EU relationship, China was committed to a long-term, stable, and full partnership with the EU with the following objectives:

1   To promote a sound and steady development of China–EU political relations;
2   To deepen China–EU economic cooperation and trade; and
3   To expand China–EU cultural and people-to-people exchanges.

The EU Policy Paper identified various fields such as politics, military, economy, education, science, culture, and social and judicial systems to expand the cooperation, with civil aviation being one of the industries that required more collaboration. "China-EU exchanges in civil aviation will be deepened and Chinese and EU enterprises are encouraged to strengthen their cooperation on production, technology, management and training in the aviation sector" (FMPRC 2003). The paper called on both sides to make efforts to improve economic relations over the next five years, such as to give play to the mechanism of an economic and trade regulatory joint committee and a stepped-up economic and trade regulatory policy dialogue in order to achieve the objectives of elevating the EU into China's largest trading and investment partner.

China issued its second policy paper on the China–EU relationship in 2014 declaring it welcomed further cooperation with the various EU legislative bodies and its various political parties. In respect of aviation specifically, the statement reinforced China's commitment to working closely with the EU to advance cooperation on civil aviation alluding specifically to cooperation in areas of production, technology, management and training. All in all, China has given the same importance to its relationship with the EU to that of China–USA, thus facilitating the development of a strategic relationship between the two parties.

The European Union's influence on China's air transport policy and on individual member state's negotiation

The EU's treaty making power in bilateral negotiations

Until 2002, the sovereign governments of individual member states were still negotiating their own bilateral ASAs with third parties outside the EU. In the late 1990s, quite a few EU member states including Denmark, Sweden, Finland, Belgium, Luxemburg, Austria, and Germany had concluded or had been negotiating with the USA for an open skies agreement. This practice was criticised by the commission which complained that the individual arrangements between the member states and the USA did not reflect the EU's core values in that it did not protect the interest of the EC as a whole. Being authorised by the EC Treaty to conduct international trade negotiations, the commission maintained that negotiations by individual member states would jeopardise the potential benefits available to those that had not concluded open skies deals with the USA.

Consequently, the commission took a number of EU member states to the European Court of Justice (ECJ) in December 1998, accusing them of having breached their duties under the EC Treaty based on the grounds that only the European Commission had the treaty-making power with a third party outside the EU. To the extent that it was necessary for the EU to implement the treaty and satisfy its objectives, to ensure fair competition, and to preserve the established community laws, entrusting the EU with this power was deemed necessary. It stressed that the nationality clauses in the existing ASAs would prevent community carriers from benefiting from the community market by depriving them of the right to establishment, thus preventing consumers from gaining the benefits of increased competition.

In November 2002, the ECJ ruled against the member states, pointing out that in accordance with the community law, the agreements signed between those member states and the USA infringed the rights of community carriers to non-discriminatory market access to routes between all member states and third party countries. The ruling was intended to entrench the EUs freedom of establishment principle within the aviation industry, but the opening of European skies to US carriers was not reciprocal for all EU carriers. Although the member states did not infringe on the treaty-making power of the community based on Regulation 95/93, some of them infringed their obligations according to Article 10 EC and Regulations 2409/92 and 2299/89 by concluding the open skies agreements with the USA as these agreements contained regulations concerning the applicable fares on intra-community routes and the use of Computer Reservation System (CRS) (Bartlik, 2007). It was these subject matters that fell into the exclusive treaty-making power of the European community. The court also found that the national ownership and control requirements in those bilaterals cited by the EU actually violated a central principle of the European Treaty (Williams, 2009). The causal factor was found to be located in the designation process, where the nominated carriers were either owned or controlled by the signatory member states, which had a serious effect in law because it denied the rights of airlines in other member states to receive any form of national treatment after the ASA was duly signed and sealed (Williams, 2009). The court further identified that an additional 1,500 existing agreements signed with foreign countries other than the USA were in violation of the treaty and measures must be taken to amend these ASAs to bring them in conformity with EU law.

To resolve the discrepancy, it was concluded that agreements between the member states and their ASA partner countries could be amended and negotiated at bilateral levels, or the EU could negotiate single "horizontal agreements," with the commission acting on behalf of the member states. The consequence of the ECJ ruling cannot be underestimated. It was a milestone marking the start of a community external aviation policy and gave rise to the commission being entrusted with exclusive responsibilities in aviation policy with third states. Though member states were still entitled to negotiate bilaterally with any third country outside the EU, those negotiations had to ensure there was conformity with the established EU law. A mechanism has since been established where the member states must coordinate, update, and consult with the EC the progress of any negotiations should they engage themselves with any third country (Geil, 2010).

The incumbent method, the horizontal agreement, does not replace the existing bilateral ASAs or impinge on the traffic rights afforded. Instead, it makes amendments to relevant clauses, such as the nationality clause, to ensure community carriers enjoy non-discriminatory market access to third states. Some key countries were identified including China, Canada, the USA, Japan, Australia, and New Zealand, with which the European Commission wished to start negotiations, although it had not necessarily received mandates for such negotiations. However, interests in these and other markets grew—undoubtedly driven by the 2015 aviation strategy where the council sought the authorisation of member states to implement EU-level agreements with various countries and regions. The historical open skies deals (Stages I and II) concluded with the USA in 2007 and 2010, respectively, were one of such horizontal agreements negotiated by the commission with a mandate received from the council. In 2011, there were further ascensions to the EU–USA air transport agreement by way of the integration (ascension) of Norway and Ireland. The EU–USA open skies agreement is considered a model for other open skies agreements because it has tackled the "nationality clause," "ownership," and "cabotage" issues, which have remained unchallenged in any of the previous open skies agreements between the USA and its counterpart signatories, although they are not a problem within the EU which is the first fully liberalised region. By 2019, the negotiations at the bilateral level had led to changes with 73 partner states, accounting for 340 bilateral agreements. At the community level, horizontal negotiations resulted in changes with 41 partner states and one regional organisation which encompasses eight member states, representing 670 additional bilateral agreements (European Commission, 2019).

EU's attempts to negotiate a horizontal agreement with China

Following the open skies judgment of the ECJ in November 2002, the commission had been keen to negotiate a horizontal agreement with China, which was identified as one of the key countries to resolve the legal issue. To test how the Chinese would react to such a request, the commission sent a senior delegation in May 2004 to explore the possibility of negotiating at the community level to amend the existing ASAs concluded between the member states and China. Encouraged by CAAC's positive feedback, in March 2005, the commission requested the council to authorise a mandate so as to allow it to proceed with a comprehensive OSA with China, arguing that it was time to develop a wider range of opportunities between China and the community in air transport

(Commission of the European Communities 2005). A comprehensive mandate was imperative, paramount, and mutually beneficial given:

i the growing importance of the Chinese aviation market;
ii the difficulties encountered by individual member states in seeking to bring bilateral agreements into conformity with community law; and
iii the benefits of replacing the fragmented European approach in its relations with China with a coordinated and liberal approach.

The commission reiterated that a mandate for negotiating with China would be adequate to reflect the EU's position in the world today and clearly strengthen the negotiating position of the community to signal its determination to resolve the outstanding infringements of community law and competence. Supported by the findings of research, the commission argued that the benefits of opening market access and easing investment rules in China would be similar to that of an EU–USA OAA which was able to generate several hundred millions of euro business per annum.

To convince the council that the commission was more competent in negotiating with China, it further spelled out the objectives to be achieved as well as the strategies to be employed in conducting such negotiations. The amended agreement with China, as the commission suggested, should not be limited to agreeing on community designation clauses, but establishing an ambitious framework integrating industrial cooperation and wider aviation issues such as cooperation in the fields of aviation safety, security, air traffic management, technology, and research, as well as "doing-business" issues (Commission of the European Communities, 2005). Such a comprehensive scope would enable both parties not only to address the legal issues to ensure a level playing field but also to explore opportunities to foster technical and industrial-wide cooperation (Commission of the European Communities, 2005).

To ensure that the Chinese side was fully aware of the significance of the amendment of the existing ASAs between EU member states and China, a second exploratory meeting was held in Beijing in October the same year with the EU delegation providing a general overview of the development on the negotiation and implementation of the existing ASAs between the EU member states and third countries. It made it clear to the Chinese side that there was an urgent need to amend such existing ASAs between China and EU member states to include a community designation clause, without which such agreements would remain vulnerable to legal challenges (Commission of the European Communities, 2005). China replied with caution that internal assessments had been undertaken based on the possibility of solving the outstanding legal issues as clarifications were being made in relation to the draft of the horizontal agreement. Instead of committing to engaging in such negotiation with the community immediately, China only promised to give further consideration to such an agreement.

Further pressure was exerted at the seventh EU–China Summit held in the Hague on 8 December 2004 which was attended by top senior officials including Chinese Premier Wen Jiabao and his team of ministers as well as the President of the European Council, and the Prime Minister of the Netherlands, Mr. Jan Peter Balkenende and his team. Believing that the summit was the most appropriate venue to inform China of EU's determination to resolve the legal issues in the existing bilateral ASAs, the commission brought up the issue once again and requested that it should be treated as a priority to bring them in conformity with EU law. The call received positive feedback from the Chinese counterpart with a commitment to addressing the issue being recorded in the joint statement released on 8 December 2004, stating that

> the Leaders, in welcoming the dialogue progress and the closer EU-China co-operation in civil aviation, agreed to further deepen co-operation in this field, including civil aviation industry and technology, and to solve as a matter of priority outstanding issues thereby opening the way for a potential future EU-China Aviation Agreement.
>
> *(Council of the European Union, 2004)*

The EU won China's commitment to engaging, as a matter of priority, in formal negotiations of the "horizontal agreement" in the following June, when the first EU–China Aviation Summit was staged in Beijing sponsored by the EU and jointly organised with CAAC. Gathering more than 200 senior government officials including Mr. McMillan, director general of civil aviation of the UK acting as the EU Presidency at the summit, and industry professionals, the summit was considered the best mechanism to draw attention to the necessity for a horizontal agreement at the community level. In addressing the opening ceremony, Mr. Jacques Barrot, vice president of the commission, raised the issue once again by referring to the commitment made by both parties at 2004's seventh EU–China Summit, stressing that it was one of the priorities of the EU to ensure that aviation relations were founded on a sound legal basis. Those ASAs with third countries including China that were not in conformity with community law must be amended to provide legal certainty to all operating airlines (Barrot, 2005). He told the summit that the Council of European Transport Ministers had adopted conclusions in which they agreed on a roadmap for Europe's external aviation policy for the coming years and urged his Chinese counterpart to "take a favourable decision to engage in the exercise shortly" to resolve the issue so that more time and energy could be dedicated to negotiations of a broader, global EU–China agreement (Barrot, 2005). At the meeting between himself and the then CAAC Administrator Yang Yuanyuan on 29th June 2005, the day prior to the official opening of the summit, both sides expressed the wish to strengthen cooperation, notably through three concrete initiatives, one of which was to take action to bring those existing bilateral agreements between EU member states and China in line with the EU law.

The following years saw continued communications and negotiations between the EU and CAAC, although progress was slow. This was on the one hand because of China's concern about the prospective consequences that might result from the implementation of the amended designation clause (Li, 2009), while on the other hand, the commission was also aware of the challenges it would have as a consequence of such a comprehensive agreement with China. It fully understood that restrictions on traffic rights might remain under a future community–China agreement with the need to develop transparent and non-discriminatory procedures for the allocation of such limited traffic rights as negotiated at the community level (Commission of the European Communities, 2005). Such a commitment to preparing the procedures of such an allocation and its gradual implementation was not adequate to persuade its Chinese counterpart that the mechanism would not be discriminative to Chinese carriers that were struggling in the China–EU market.

Encouragingly, continuous dialogues between the two parties brought about a consensus in 2009 with the commission agreeing with CAAC to continue bilateral negotiations with selected EU member states to resolve the legal uncertainty under the precondition that a "circumvention clause" is incorporated in the bilateral agreement (Liang, 2010). The agreement between China and the Netherlands reached in March 2010 was a good showcase to demonstrate that with the continuous communication between China and the EU, its member state was able to resolve the legal issue while at the same time relaxing the market-entry restrictions through bilateral negotiations. By October 2010, successful negotiations were completed with Italy, Belgium, Norway, Sweden, Denmark, and Slovakia, and the EU's requirements for designation and nationality were accommodated. These developments served as a precursor to the previously mentioned letter of intent signed in 2013 to enhance cooperation in aviation between China and the EU. To extend the partnership in aviation, an EU–China Aviation Partnership Project (APP) was subsequently established in 2015 by the European Union Aviation Safety Agency (EASA) and CAAC to the tune of ten million euros and was commissioned to run from 2015 to 2020. The APP was defined as having eight major areas of cooperation including:

1 Regulatory dialogue on safety,
2 General aviation,
3 ATM/ANS and airports,
4 Airworthiness,
5 Environmental protection,
6 Economic policy and regulation,
7 Aviation safety and security, and
8 Legislation and law enforcement.

It was established through the partnership of the EU chamber of commerce, aviation authorities of the member states, the China Academy of Civil Aviation

Science and Technology, and the Air Traffic Management Bureau of Civil Aviation Administration of China.

Impacts on China–UK and China–Netherlands's negotiation

Institutions such as the European Commission at the EU level are supranational organisations that are authorised by its member states to act on their behalf in certain realms. Such a regime is quite unique in the international system and China, although comfortable in dealing with it, is not comfortable in negotiating with it any kind of commercial arrangements, with one of the biggest concerns being the consequences of treating divergent social and economic systems as one. Holding that bilateral negotiation with an individual country for a country-pair market is the best mechanism to address international air transport issues, China needs to be certain that any arrangements between the country and the community will not jeopardise its national interest. Whenever possible, the Chinese would avoid negotiating with European interlocutors who are empowered to speak for the EU as a whole, which brought its combined weight to bear (Fox and Godement, 2009). China prefers dealing with national negotiators, a tactical choice that many member states also preferred, which allowed them to exert national control (Fox and Godement, 2009).

From the EU's perspective, its priority over the years since 2002 has been to resolve legal issues in the existing bilateral ASAs. Without a mandate from the council of ministers, the EU was not eligible to negotiate such a comprehensive agreement with China, which is more than happy to deal with individual EU member states. Although EU has developed a policy outlining its goals and objectives in dealing with China, as well as formulated a regime of summits gathering the state heads of its member states and China on a regular basis, it has failed to produce any substantial consequences to the benefit of the EU because the objectives are seldom followed through due to the steady increase in the number of objectives and competing requirements for urgency (Fox and Godement 2009). The EU's pressure on specific issues tends to come through formal dialogues, which China accepts and is engaged in continuation, but unfortunately lead to inconclusive talking shops (Fox and Godement, 2009).

In addition, EU member states have differing philosophies about how to deal with China despite both the Netherlands and the UK advocating a liberal unrestrictive trade regime (Fox and Godement, 2009). The differences between member states, while serious, are nowhere near as significant as that between any one of them and China (Fox and Godement, 2009). Their divergent views and me-first strategy in dealing with China has jeopardised a unified EU approach, which has undermined the EU's position in negotiating with China (Fox and Godement, 2009).

Traditionally focusing on international trade and commerce, both the Netherlands and the UK take international air transport relations with China as economic issues and are reluctant in bringing these into the political realm (Knight, 2010).

Aviation including international air transport is only a subject that is covered in industry cooperation which does not receive high-profile attention at the summit level. Evidence from the empirical research in the preceding chapters has revealed that neither party has involved any senior officials above the ministerial level in the bilateral negotiations, nor are any political figures or mechanism involved in such bilateral negotiations. The respective competent authorities take full responsibility to negotiate at the ministerial level, which has proved to be effective and successful in addressing market demand to support bilateral trade.

References

Barrot J., (2005), Open speech at the EU-China civil aviation summit, Beijing 30th June 2005, available at europa.eu/rapid/pressReleasesAction.do?reference=SPEECH/...

Bartlik, M., (2007), The impact of EU law on the regulation of international air transportation, Ashgate Publishing Limited, Aldershot.

Casarini N., (2006), *The evolution of the EU-China relationship: from constructive engagement to strategic partnership*, The European Union Institute for Security Studies, Paris.

Clinton H. R., (2009), Closing remarks for US-China strategic and economic dialogue, Washington, DC, 28 July, 2009.

Cohen S. D., Blecker R. A., and Whiney P. D., (2003), Fundamentals of US foreign trade policy, Economics, politics, laws and issues, (second edition), Westview Press, Boulder, CO and Oxford.

Commission of the European Communities, (2005), Communication from the commission: A community civil aviation policy towards the People's Republic of China- strengthening cooperation and opening markets, Brussels, 11. 03.2005, COM(2005)78 final.

Council of the European Union, (2004), Joint Statement of the 7th EU-China Summit, the Hague, 8th December 2004, Brussels, 15065/04 (Presse 337), available at www.consilium.europa.eu/uedocs/cms_data/docs/pressdata/en/er/82998.pdf

Council of the European Union, (2006), "EU-China: closer partnership, growing responsibilities", Communication to the Council and the European Parliament, COM (2006) 631 final, Brussels, 24 October 2006, p. 2, available at http://ec.europa.eu/external_relations/china/doc/06-10-24_final_com.pdf

Council on Foreign Relations, (2007), China Strategy Paper, available at: www.cfr.org

Diao Y., and Li X., (2008), High-level dialogue vital to ties, 06 December, 2008, China Daily, Available at www.chinadaily.com.cn/bizchina/2008-12/06/content_7279045.htm

Dinan D., (1999), *Ever closer union, an introduction to European integration*, Lynne Rienner Publishers, Inc, London.

European Commission., (2019), External aviation policy - Horizontal agreements, available at https://ec.europa.eu/transport/modes/air/international_aviation/external_aviation_policy/horizontal_agreements_en

Feng H., (2006), *The politics of China's accession to the World Trade Organization*, Routledge, London and New York.

Friedman, M., (1995), Foreign economic aid: Means and objectives, No. 60, Hoover Press, Stanford.

Fox J., and Godement F., (2009), *A power audit of EU-China relations*, London: The European Council on Foreign Relations, Policy report, available at http://ecfr.3cdn.net/532cd91d0b5c9699ad_ozm6b9bz4.pdf

Garrison J., (2005), Understanding prospects for cooperation and competition in US-China relations-a review of theory and practice. Paper prepared for the annual meeting of the International Studies Association, Honolulu, Hawaii, USA.

Garrison J. A., (2007), Managing the US-China foreign economic dialogue: building greater coordination and new habits of consultation, in Asia Policy, Number 4 (July 2007), 165–185.

Geil K., (2010), Interview conducted in April 2010 at the EU head office in Brussels.

He K., (2009), Dynamic balancing: China's balancing strategies towards the United States, 1949–2005. *Journal of Contemporary China*, 18(58), 113–136.

Khalilzad Z., (1999), The United States and a rising China: strategic and military implications, Washington, RAND, available at www.rand.org

Knight S, (2010), Telephone interview conducted in December 2010 in the UK.

Lampton D. M., (2001), *Same bed different dreams, managing US-China relations 1989–2000*, University of California Press Ltd., Berkeley, Los Angeles and London.

Lancaster C., (2007), *Foreign aid: diplomacy, development, domestic politics*, The University of Chicago Press, Ltd, Chicago, IL.

Li J. X., (2009). The world civil air transport and the growth of China's civil air transport. available at www.caac.gov.cn

Liang N., (2010), Interview conducted in Beijing October 2010.

Men J., (2007), The EU-China strategic partnership: achievements and challenges, Policy paper No. 12, Nov. 2007, University of Pittsburgh, University Centre for International Studies, EU Centre of Excellence/ESC Website available at www.ucis.pitt.edu/euce

FMPRC, (2003), China's EU policy paper, available at: https://www.fmprc.gov.cn/mfa_eng/topics_665678/ceupp_665916/t27708.shtml

Paulson H. M. Jr., (2008), The right way to engage China, strengthening US-China ties, *Foreign Affairs*, Sept/Oct Issue.

Radelet S., (2006), A primer on foreign aid, Working Paper Number 92, July 2006, Centre for Global Development, available at www.who.int

Reimold D. B., (2009), "Airports and the environment", Opening Remarks at China-US Aviation Symposium (8th April, 2009).

Rogin J., (2010), Does the strategic and economic dialogue matter? available at www.thecable.foreignpolicy.com on 21 May 2010.

Shi T. with Wen M. C., (2009), Avoiding mutual misunderstanding: Sino-US relations and the new administration, January 2009, Carnegie Endowment for International Peace, Foreign Policy Summary.

Tymczyszyn J., and Bai Y., (2009), ACP making a difference: model public-private partnership: AmCham-China's US-China aviation cooperation program creates big benefits, 04 August 2009, available at www.amchamchina.org

USDTA, (2009), USTDA launches environmental partnership at China-US aviation symposium, reaffirms commitment to US-China aviation cooperation program, USTDA press release on 8th April, available at www.ustda.gov.

Wai J., (2005), EU-China relations: economic, political and social aspects; Department of Government & International Studies; Hong Kong Baptist University, available at www.soc.nii.ac.jp/eusa-japan/download/eusa…/paper_TingWai.pdf

Williams A., (2009), *Contemporary issues shaping China's civil aviation policy, balancing international with domestic priorities*, Ashgate Publishing Ltd, Farnham.

Zhang L., and Zhuang Y., (2005), Sino-EU relations in retrospect and prospect, *Foreign Affairs Journal* (Waijiao Shiwu, Beijing), No.76, 2005, available at www.irchina.org/en/news/view.asp?id=346

9

APPLYING THE MICRO-MACRO LINKAGE APPROACH (PART III)

Analysis of factors at domestic and institutional levels

Factors at the domestic level and their impact on policymaking

To fully understand the evolution of China's international air transport policy, the examination has to extend beyond the external impetus, as any such policy is formulated by a sovereign state, which must address its domestic political and economic needs. Innenpolitik theories argue that a fundamental objective of a nation's foreign policy is to ensure the survival of the political regime and leadership (Lai, 2010). Specifically, a nation's foreign policy needs to take the following into account:

- Preserving the political regime;
- Addressing economic demands which is integral to its economic system,
- Top leaders' vision for a preferred political regime, political survival, and their skills in managing the external–internal relationship (Lai, 2010).

Whether the policymaking power is centralised or decentralised, has a significant impact on the policymaking process, thus varying the policy outcome. A centralised policymaking mechanism is able to coordinate conflicting interests and dissenting views with relative ease, so as to allow the central policymaking institution to formulate coherent foreign policies and conduct effective trade negotiations without serious impediment (Aggarwal and Lee, 2011).

Societal forces are another source of influence on a country's foreign policy. Hao (2005), in investigating China's foreign policy towards the USA, argued that societal forces are almost a catch-all notion embracing a variety of elements, which include but are not limited to public opinions, the business community, think tanks, media, technocrats within the bureaucratic apparatus, local governments, and other subnational entities. These interest groups would either argue for or against a policy for their self-interest and benefits (Aggarwal and Lee, 2011). Policymakers are thus constrained with the considerations of conflicting interest of the business community, with the eventual policy outcome being a trade-off. While these forces operate as a loose conglomeration of individual actors who barely coordinate with each other, together, they have the power to influence the policy outcome, directly or indirectly, intentionally or unintentionally (Hao and Su, 2005).

Air transport as an instrument to support national political and economic objectives

Since 1978, the key objective for the Chinese government has been to pursue economic growth to achieve modernisation. In its continuous search for effective strategies for economic prosperity, the government calls for industries to maximise their capacity to serve national interests. The need to develop a comprehensive multimode national transportation system including rail, road, waterway, air, and pipelines has been a strategic focus of the economic activities since 1978. The Transport system is planned more as an infrastructure rather than a service to facilitate economic growth and support economic activities (Williams, 2009). Demand for domestic and international air transport emerged in response to the increasing economic activities, which called for a more efficient multimode transport system (Williams, 2009). When pursuing export volume was the primary driving force for GDP growth, development of an efficient air transport system became an imperative task for the country's aviation industry policymaker.

Throughout the 1980s and 1990s, one of the focuses of China's civil aviation policy objectives was to nurture air transport development to facilitate foreign direct investment to boost local economic growth. This was achieved by allowing the establishment of new airlines, opening new routes, and increasing capacity. An increasing number of AOCs were issued, resulting in a handful of new airlines being intercepted, for example, Xiamen Airlines set up in 1984, Shanghai Airlines in 1985, Shenzhen Airlines in 1992, and Hainan Airlines in 1993. These airlines were registered in Xiamen, Shanghai, Shenzhen, and Haikou, respectively-the forerunners embracing economic reform and favourite destinations of international visitors and foreign direct investment. These airlines were funded by their respective municipal or provincial governments; thus, they were financially healthy to acquire the state-of-the-art aircraft and provided the best service.

After almost a decade of operations, Xiamen Airlines became the first local government-sponsored carrier to be licensed for regional operations, expanding its services to destinations in Southeast Asian countries, although under the name of China Southern. Shanghai and Shenzhen Airlines, however, were not authorised to operate in Southeast and/or East Asian destinations, such as Singapore, Thailand, South Korea, and Japan until the 2000s.

To accommodate traffic demand, airport expansion or new construction projects were approved. Thirteen airports, in addition to Beijing, Shanghai, and Guangzhou, along the east and south coastline, including Dalian, Tianjin, Qingdao, Yantai, Jinan, Nanjing, Hangzhou, Ningbo, Xiamen, Fuzhou, Shantou, Shenzhen, Haikou, and Sanya were granted for international operations, although mainly to the destinations in East and Southeast Asian countries. These international links have significantly improved the connectivity between China and the Southeast Asia.

In the 21st century, the central government announced another strategic move, calling upon all industries to render their support to the endeavours to redevelop Central and Western China and to revitalise the Northeast provinces. The objective was to minimise the imbalance of wealth accumulation between the East and West, and South and North. The South-Eastern region along the coast line was far too developed, while the North-Western regions remained under-developed, resulting in one country with many markets (Meyer, 2008). Economic prosperity and improved living standards have been largely enjoyed by coastal residents, especially along the physical parameters of the Beijing–Shanghai–Guangzhou triangle. Recognising the divergence in economic growth and the ever-growing gap within the country, the central government established the urgent need to address the imbalance to stabilise the social order, which would enhance China's image as a growing international player. To this end, several policy initiatives, such as the Great Western Development Strategy and the Northeast Regeneration, had been launched in attempts to drive economic growth in these regions. An underlying geopolitical implication was to protect the political integrity of China's national boundaries, where the majority of the population has divergent cultural practices and religious beliefs (Williams, 2009).

In response to the central government's call, China's aviation policymaker issued a series of industry-specific policy initiatives, which outlined the blueprint to assist the achievement of a redevelopment goal. Priority would be given to new airport construction and expansion projects in these regions and direct international airlinks would be pursued. For example, between 2005 and 2010, a total of $6.5 billion would be invested in airport construction and expansion in Central and Western China. Kunming and Chengdu Airports would be enhanced to become regional hubs connecting China and Southeast Asian countries. Four new airports in addition to Lhasa Gonggar airport would be built in Tibet: Qamdo Bamda, Nyingchi, Gunsa, and Xigaze airports. By 2020, a total of 250 airports are expected to be in service (State Council, 2017). MOUs were signed between the CAAC and provincial governments such as Xinjiang, Yunnan, Jiangxi, and Henan, with the CAAC pledging to support the provincial efforts in developing more direct international links.

In alignment with the central government's "Going Global" initiative, designed to encourage Chinese enterprises to invest overseas, in contrast to the open-door policy that invited foreign capital into the Chinese market (Ku, 2006), the CAAC negotiated liberalised arrangements with its African and Southeast-Asian counterparts and encouraged its airlines to provide direct links to those destinations. For example, restrictions on cargo operations on Sino-African markets were removed and the seventh freedom traffic rights operations on passenger/combination services were allowed. Restrictions on the third, fourth, and fifth traffic rights passenger/combination operation on selected China-ASEAN country-pair markets were also relaxed, allowing multiple designations including LCCs to enter the Chinese market.

Similar testimony can also be found in the Belt and Road Initiative (BRI). The evolution of a proverbial "air silk road" is the ambit of the Civil Aviation Authority of China which is being strategically executed to support the mission of the BRI. While the Chinese carriers were encouraged to aggressively develop airlinks, the authority has also been positioning itself to ensure the framework is instituted for the expansion to occur. In the 13th Civil Aviation Development Five Year plan, 74 new civil aviation airports were outlined to be commissioned by 2020 drastically expanding access across mainland China and creating nodes in areas with low airport density. The expansion of these nodes also creates expansion opportunities for cities which have not traditionally been vibrant economic centres. Henan's Zhengzhou city is located approximately two hours from the three principal industrial zones across China which collectively account for 95 percent of China's GDP. The BRI magnifies Zhengzhou airport's bid to solidify its position as a logistics hub due to its geographical location. As an industry still in the throes of deregulation, low connectivity, and limited infrastructural developments in some BRI partner countries, aviation's role in the BRI, although not unencumbered, is still laden with opportunities.

The BRI has been strengthened in some respects due to the input of the aviation industry. It is an opportunity to develop the civil aviation industries of the developing countries along the route which suffer the same challenges such as personnel shortages, financing, infrastructural, and technical challenges. The CAAC reported that China has thus far signed air transportation agreements with 125 countries and regions, of which 62 are participants in the BRI and with direct air services to 45 of these, totalling 5,100 flights weekly (Xinhua, 2018). China's investment in aviation industries among belt and road countries can be expected to fuel the growth and further internationalisation of China's aerospace industry. According to the CAAC, there are strategic projects concerning aviation in the BRI amounting to an investment of RMB 200 billion (Ernst & Young, 2016). As a consequence of the investment and expansion of bilateral arrangements, passengers throughput continued to increase relative to previous years. Most notably, in 2015, passenger flow bettered the previous years by 70 percent (Ernst & Young, 2016).

The BRI has perpetuated rapid investments by Chinese companies in aviation infrastructure in partner countries such as airports. Investments by Chinese firms in foreign airports in Germany, UK, Italy, France, and Albania are supplemented by Chinese impetus towards breaking the extant duopoly in aircraft manufacturing between Boeing and Airbus. To this end, Chinese aircraft manufacturer Comac produced a narrow body jet "C919," with first delivery expected in 2021. The aircraft is slated to offer competition to similar narrow body jets (Boeing 737 and Airbus A320), and signals Comac's foray into aircraft manufacturing on a globally competitive scale. A new entrant in a field where barriers to entry can traditionally be extensive, there are reportedly approximately 800 orders for the aircraft from primarily Chinese and other Asian airline operators.

In cases where other manufactures have developed technologies capable of challenging the market dominance of the existing duopoly, their responses to the competitive threat have either sought to extend their market power and thus their market dominance, or barriers to entry were heightened. The latter was evident insofar as regulatory barriers were raised when Canadian manufacturer Bombardier developed a narrow body twin engine mid-range aircraft designed to compete in the market with similar products such as the Boeing 737. However, the US-based Boeing argued that the Canadian government's provision of subsidies to Bombardier enabled the manufacturer to produce and sell the aircraft at a price substantially lower than its fair market value. In response, the US government sought to apply tariffs of approximately 300 percent, although it was subsequently reversed after a ruling by the US International Trade Commission overturned the decision in Bombardier's favour. Ultimately, Airbus purchased a majority stake in the C series bringing its market dominance to bear.

Comac, still in a nascent stage of development, has not frontally challenged the duopolistic environment as the manufacturer is yet to deliver its first order. However, the growth of the Chinese aviation market, the rapid expansion of Chinese carriers through China's increasing outbound travel, and its rapid tourism growth create unique opportunities for Comac to wrestle market share from the incumbents. Aviation being considered to be one of the ten primary pillars of the "Made in China 2025" policy initiative which seeks to further expand China's dominance in advanced manufacturing, even further magnifies the potential for Chinese aircraft manufacturers.

Cliff et al. (2011) recount that as early as the 1980s without the domestic manufacturing capability to supply China's aviation requirements, the Chinese government had to permit local airlines to import such aircraft. However, the need to revolutionise the domestic aircraft manufacturing industry and to make it globally competitive resulted in manufacturers being encouraged to establishing joint ventures with state-owned aerospace corporations ultimately facilitating technology transfer and tacit knowledge. The strategy was somewhat limited in its effectiveness as much of the modern technology was still being developed in the west and shipped to China for assembly. Over time, the joint ventures established with Chinese manufacturers, coupled with partnerships with western suppliers in the value chain enabled the development of the necessary tacit knowledge and access to technology to strengthen China's position as a credible base for aircraft manufacturing.

BRI creates opportunities for aviation activity fuelled by China beyond an increased propensity for Chinese manufactured aircraft. As China and other BRI countries extend their socio-political and economic integration through the initiative, airline capacity between these nations can be expected to continue to increase. In 2015, the China–Pakistan economic corridor formed as one of the central pillars of the BRI in part included the financing of a new airport, that is, Gwadar International Airport, to help strategically position the area as a regional hub. Set to be constructed on 4,300 acres, at a price tag of US $230 million (CPEC, 2017), the New Gwadar Airport will become the largest airport in Pakistan and serve as a vital supplement to the wider BRI investment in the region as an integral conduit for Beijing to Africa and the Middle East. While construction is ongoing, the uptake in airlift capacity into Pakistan already commenced and is being propagated by Chinese carriers. The annual capacity on the country pair increased by 148 percent between 2012 and 2017, driven notably by capacity improvements from Air China, Pakistan International Airlines, and the entrance of a new carrier into the market.

China's strategy to build world-class state-owned-enterprises (SOEs) including airlines

SOEs are one of the key features of China's economic system. Over the years, the SOEs have remained to be the government's favourite enterprises, despite a series of cautious and explicitly experimental policy have been formulated and executed, resulting in dynamic structural changes of the SOEs. In the early days of China's reform, the government was very keen to empower the SOEs with greater autonomy so as to hold them accountable for their own financial performance. Incentives were provided and a proportion of profits were allowed to be retained. However, restructuring became a priority after 1992 with an aim to improve the poor performance of a significant number of SOEs. The Company Law passed in 1993 called for all businesses to strive for their financial well-being through measures such as reengineering their financial structures, which resulted in medium and large scale SOEs being listed on stock markets, while smaller SOEs were divested by lease or sale to individuals. Private enterprises were encouraged to contribute to the national economic growth. The structural reform changed the ownership of majority of the industries, leaving up to 70 percent of those restructured firms to be partially or fully privatised.

Despite the continuous structural reform, the central government has never lost its control over the major SOEs, believing firmly that the state should control the key sectors of the national economy. Committed to forging a socialist market economy, the government argued that the development of strong and competitive SOEs was paramount and fundamental to retain the country's socialist characteristics and keep on track its economic reform. The principle that "the fundamental role of the state-owned economy should be reflected in its ability to control the national economic system" was established in the Party's

15th Plenary Session in 1997. To this end, "the state-owned economy should be restructured strategically so that the state holds an absolutely dominant share in the key industries and critical sectors which concern the livelihood of the national economy" (Ji, 2010). The party further specified that the state-owned economy should include those natural monopoly companies and prominent businesses of high technology in such industries as defence, electricity and grid, petroleum and gas, telecommunication, coal and mining, shipping, and air transport, which have a stake in the national security and provide essential public products and services.

SASAC was thus created in April 2003 under the direct control of the SC, whose primary task was to execute the initiative and oversee the restructuring of SOEs for its long-term competitiveness (Ji, 2010). In 1997, the 500 largest state-owned firms only held 37 percent of the state's industrial assets, contributing 46 percent of all tax revenues from the state sector (China Labour Bulletin, 2007). In 2009, all of China's central SOEs generated 12.6 trillion Chinese yuan of revenues in sales, realising 815 billion Chinese yuan in profit (Li and Brødsgaard, 2013). The concentration of the core business of SOEs has been in major industries such as telecommunications, petroleum, defence, and aviation where the World Bank (2012) estimates that the SOE operations dominate.

The Chinese airlines industry has been instrumental in supporting the national economic growth. It also has a substantial stake in the nation's security interest, with state ownership being the dominant theme. By deliberately classifying airlines into two categories, namely the administration-affiliated and non-affiliated carriers, different treatment could be justified to different categories of airlines through regulatory measures. The administration-affiliated carriers were the six carriers that were spun off from the CAAC between 1987 and 1990, while the non-affiliated ones were the joint ventures set up by the municipal/provincial governments and local businesses such as Xiamen Airlines (1984), Shanghai Airlines (1985), and Shenzhen Airlines (1992). For example, in the late 1980s and early 1990s, the CAAC would only allow three of its affiliated airlines to operate on international routes, with the rest being allowed to serve domestic trunk routes between Beijing, Shanghai, and Guangzhou, as well as other major routes. For the non-affiliated carriers, approval was only given to provide domestic services connecting major airports from their home bases respectively or other feeder routes. Shanghai Airline was an exception in serving Shanghai and Beijing due to its association with Jiang Zemin and Zhu Rongji, who used to be the city's party secretary and mayor before taking the role of state head and premier. The non-affiliated airlines, such as Xiamen Airlines, did not receive its first international route licence for regional and peripheral operations until January 1996, 12 years after its initial launch (XiamenAir, 1996), while Shenzhen Airlines was only licensed on international routes in 2008, after almost 16 years of operations. In controlling fleet planning, the Chinese regulator was able to allocate medium to long-haul wide-body aircraft to its affiliated airlines so as to support their trunk and international routes expansion, while leaving short-haul and feeder aircraft to the non-affiliated airlines.

As the reform intensified, affiliated carriers were encouraged to raise capital in both domestic and international stock markets, with China Southern and China Eastern being listed in the Shanghai, Hong Kong, and New York markets in 1997, and Air China in the Shanghai, Hong Kong and London markets in December 2004. Air China made available 1.12 billion shares in the Hong Kong market, representing 40 percent of its total global offering of 2.8 billion shares. In contrast, when Hainan Airlines (1993) was listed in the domestic stock market in 1993, the Chinese regulator was not impressed with its behaviour although it failed to take any action since the carrier was only a non-affiliated operator, over whom the regulator was only authorised to exert economic and safety regulation but not financial and human resource control (Le, 1997). When in 1994 the government of Hainan Province sold 25 percent of its shares in Hainan Airline to a US investor without consulting the regulator in advance, the agency was angered by the move denying that the foreign investment was allowed in core aviation business (Le, 1997). As a consequence, the airline had to rectify its transaction, resulting in the deal going through as an investment in the airline's CRS (Le, 1997). The airline had not received its first international route licence until February 2004, when it signed an agreement with Hungarian Airlines (Malev) to launch a Beijing–Budapest service, with conditions that the carrier must launch the service no later than September the same year.

As increasing numbers of Chinese carriers have been authorised for international expansion, the central government held the view that stronger and more competitive state-owned carriers should be nurtured to their full potential. The three megacarriers were thus brought under the wing of SASAC in 2002 so that effective support and protection could be provided in the course of their growth where needed. SASAC was created as the watchdog of the central government (Kong, 2010), whose mission was to carry out the state's role as owner in the industrial economy. Although floatation was allowed for some of the companies in key industries in both domestic and international stock markets, the central government upheld the philosophy that state should maintain substantial control of those SOEs (Ernst and Naughton, 2008). Believing that SOEs have the capability of maintaining a monopoly position in the market, the government reinforced the importance of focus on core business, their financial returns, and performance rather than economic growth, reform, and fair competition (Ernst and Naughton, 2008).

While managing the assets of 123[1] SOEs on behalf of the nation worth \$3.1 trillion (21 trillion Chinese yuan), SASAC was also responsible for recruiting and appointing the senior management positions of those SOEs. If required, SASAC would offer state aid to combat difficult times. Since December 2008, China Eastern has received \$1.3 billion cash injection, China Southern was granted

$411 million, and Air China pocketed $220 million, which were used to save them from bankruptcy, cut debts, and pay back their loans (Cantle, 2010). In June 2009, SASAC brokered the merger between China Eastern and Shanghai Airlines, with an objective of optimising resources in the Shanghai market so as to develop the city into a world-class multi-modal hub to compete with other cities in the region.

It is not surprising that the government-backed carriers lived well through the unprecedented financial crises. For the first half year of 2010, all three megacarriers reported a magnificent rise in operating revenues, with Air China recording 34.3 billion Chinese yuan ($5.04 billion US dollars), an increase of 53 percent over 2009; China Eastern recording 33.6 billion Chinese yuan ($4.94 billion US dollars) with a 92 percent increase; and China Southern recording 34.7 billion Chinese yuan ($5.10 US dollars), a 39 percent increase over the previous year. Figure 9.1 compares the net profit of the three carriers for the first half year from 2008 to 2010.

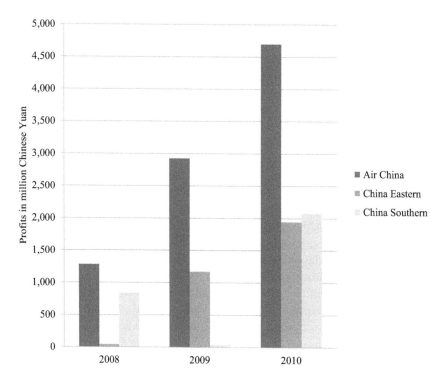

FIGURE 9.1 Comparison of net profits of the three mega-carriers for the first half year from 2008 to 2010 (in million Chinese Yuan).

Source: Individual airlines annual report in 2008, 2009, and 2010.

Latest development: a harmonised society and economic development, China's new philosophy governing economic reform

In embarking on its economic reform since 1978, China has faced the dual tasks of dismantling socialist institutions while constructing ones that promote a market-oriented approach towards growth (Tsai and Cook, 2005). By converging towards liberal trading practices, the country is able to liberalise its trade regime in a much shorter period of time compared with other nations in the region (Pekkanen and Tsai, 2005). In this process, the Chinese government has played a pivotal role in initiating policies to transform its centrally controlled economy to a market economy.

While appreciating the success of its economic reform in improving people's life standards and enhancing the national comprehensive power, the government has also recognised some social problems accompanying its rapid economic growth. Issues such as uneven urban, rural, and regional economic development, increasing pressures from human resource conditions, employment, wealth distribution, education, health care, housing, industrial safety, crime prevention, and public and social security are prominent and have been evolving into a major threat to the country's social stability and further development, which has had a direct impact on people's life (Han, 2008). Concerns over social disparity and tensions have led to a major campaign spearheaded by the central government in 2005, which called for efforts of the entire nation to bring harmony to the society, the economic facet of which includes a better distribution of wealth to narrow the gap, increase in employment, increase in the funds for pension, unemployment benefits and health insurance, and the efforts to protect the environment. It also meant advocacy of democracy, improvement of the democratic legal system and human rights protection, promotion of people's moral standards, and securing public order (Li, 2008). The rationale was that a sustained and coordinated economic growth could only be achieved where a social equity and justice system be established and maintained (People's Daily, 2005). It was regarded as a major policy change since the reform era began in 1978 which did not describe economic growth as the overriding goal of the country (Li, 2008).

To respond to the central government's call, the aviation regulator issued a "CAAC Policy Proposal with respect to the Party's Principles of the Sixth Plenary Session of Sixteenth Congress," which spelt out the following targets to achieve by 2020, contributing to a harmonious civil aviation industry (CAAC, 2007):

- The RPK should take more than 20 percent of the nation's overall transportation modes;
- The serious accident rate per one million flying hours should be less than 0.15 percent;
- Passenger satisfaction should be higher than 90 percent;

- Flight delays should be lower than 15 percent with delaying time being controlled within 0.5 hour against the scheduled departure time;
- Economic efficiency being improved and a fair and orderly competition environment created;
- A coordinated development of air transport system between the east and the west, trunk and feeder routes, passenger and cargo operations, domestic and international, and regular public transport and general aviation.

The Policy Proposal reiterated that safety was paramount priority in the development of air transport. Overheated growth of air transport in major cities such as Beijing, Shanghai, and Guangzhou should be refrained and subject to the requirements of the improvement of an efficient national route network. Liberalisation should only be pursued progressively, orderly, and safeguarded with the objective of improving industry's overall competitiveness. As Minister Li Jiaxiang pointed out,

> China's reform and its success was achieved as a result of inspiration and dedication, aiming at optimal outcome through experiment within the environment of Chinese characteristics rather than the blind copying of foreign experiment. What is crucial is to manage the relationship between the extent of reform, the pace of growth and the acceptable level of society. The fundamental principle to guide the future development of China's air transport industry will be to advance the reform while seeking societal stability and to reinforce this stability through reform.
>
> *(Li, 2009)*

Impact on China's international air transport policymaking

International air transport was instrumental in supporting the country's political and diplomatic objectives. China firmly believed that air sovereignty is exclusive and non-invasive, thus international air links were a reflection of the diplomatic relationship between two countries. International trade were to advance the country's political and strategic interests. The industry regulator was obliged to follow the central government's policy direction in determining whether to establish air links, open the market, and liberalise the market so as to facilitate to achieve the nation's overall diplomatic and strategic objectives. Formulation of a liberalisation policy on a specific country-pair market was subject to the diplomatic conditions of the two countries, with technical issues such as market entry and capacity being determined by such considerations as to whether the carriers were eligible for national treatment and protection.

For example, back in the 1950s, China only operated in those countries with which it had diplomatic ties, such as Russia and Pakistan. Since the 1970s, more international routes have been launched to Asian countries such as Japan and Thailand. The 1980's ASA with the USA was a consequence of the establishment of diplomatic ties, while the non-ratification of Sino-Dutch ASA in the 1976 was the reaction of the Chinese government to KLM's direct service to Taiwan (part of PR China), an act regarded as a violation of China's "One China" policy.

Before 1997, when Hong Kong was handed over to the Chinese Government, the negotiations between China and the UK with respect to bilateral air transport were more related to the whole package of arrangements. The China–UK market remained conservative and constrained until 2004 in terms of market entry and capacity compared with that between China and the USA, and China and the Netherlands, which reflected the sensitive and vulnerable political relationship between the two parties, with Hong Kong being a key issue.

In the 2000s, China's international air transport has been expanded to countries in Africa and South America, which was in alignment with China's diplomatic strategic objectives in the new era. In 2006, China held 28 bilateral negotiations, 6 of which were with African countries, 2 with Asian states, and 2 with East Europe and Middle East countries. The first African route (Beijing–Dubai–Lagos) was launched by China Southern and the first South American route (Beijing–Madrid–Sao Paulo) was launched by Air China. In April 2008, an additional agreement to the China–Chile Free Trade Agreement on Service Trade was signed, which included aircraft maintenance services, sales, and marketing of air transport services, CRS services, and airport ground handling and operation services, which was in alignment with China's multi-polar diplomatic strategic objectives that the aviation sector was required to offer its support.

Influence of interest groups

Influence of industry interest groups

Industry associations as interest groups
Interest groups are generally involved both directly and indirectly in the process of policy-making. Though unlike in the US and the EU where consultation with stakeholders is a routine practice in the policymaking process, lobbying does exist in China, the methods of involvement and the approaches of influence are different, as a result of the different culture and political systems (Gao and Tian, 2006). Interest groups cooperate in lobbying for common interests but compete for individual benefits.

The industry's association, China's Air Transport Association (CATA), was only launched in 2006 by the trunk carriers including Air China Group, China Eastern Group, China Southern Group, Hainan Airline, Shanghai Airline, Xiamen Airline, Sichuan Airline, and an academic institution, China Civil Aviation University. Currently composed of 26 airlines and 23 aviation-related organisations, CATA invests most of its efforts in regulating the ticketing agencies through licencing, processing their qualifications, monitoring their performance, and executing disciplinary actions against any misconduct. Although having participated twice in Sino-US bilateral discussions, evidence in the preceding chapters revealed that these organisations were still in its early stage in representing their members to lobby the government in the best interests of their industry members. This perhaps on the one hand is due to a shortage of competent staff, while on the other hand is a reflection of the weak role played by industry associations. Unlike those in the USA and the EU, industry associations are the representatives of individual enterprises to voice the industry's concerns to the government (Kennedy, 2005), China's business and trade associations do not have much autonomy, nor the sort of influence that the US associations have.

Individual airlines/airports as interest groups
Contrary to the weak influence of the industry associations, the involvement of individual airlines in aviation policymaking has been historically close, direct, and predominant with various effective approaches. One of the reasons is that commercial arrangements of international air transport operations between the two countries are negotiated at the government level, a system started since 1946 when the USA and the UK concluded the Bermuda I bilateral ASA, with the airlines and its national government teaming up together to negotiate with their counterparts. In the case of China, the special relationship dated back to the 1950s when China's aviation regulator and airlines were the same organisation with two titles. The regulator's best interest was aligned with that of Chinese airlines' and both reflected the same chorus.

In the 1990s China's economic reform still remained incomplete, with a mix of both planning and market forces operating in the economy. This left a great deal of ambiguity with respect to the definition of the government's role and corporate responsibilities, hence, the survival of firms was largely dependent on their relationship with government institutions, which had the authority to control the financing and the distribution of key resources (Park and Luo, 2001).

The separation of corporate responsibilities from regulatory functions of the CAAC did not change the close relationship between the two entities, who maintained an arm's length relationship (Le, 1997). The CAAC not only controlled route licencing, capacity, schedules, and pricing, but also fleet planning and financing. The CAAC also had the power to appoint leadership or simply deployed its own officials to take management roles until 2004, when these powers were handed over to SASAC. It was also a common practice for the regulator to recruit senior officials from the industry. For example, Yang Yuanyuan was the Chief Pilot of China Southern before being appointed administrator in 2003. China Eastern Group's chairman and CEO Liu Shaoyong used to be the director general of the Flight Standard Department of the CAAC before moving to China Southern as its chairman followed by the China Eastern position. The party secretary of China Eastern was previously the deputy administrator of the CAAC in the early 2000s, to name just a few. Those in the middle-to-senior management positions who rotated their roles between the regulator and airlines were countless.

One phenomenon worth noting was that such intimate relationship only existed between the industry regulator and its affiliated firms that are state-owned and under its or SASAC's direct control. Since more than 90 percent of the industry has been dominated by SOEs, it has been these SOEs that have access to the regulator and its officials to get their voice heard.

The swap of positions and relocation of senior officials within the industry have enabled an intimate relationship to be developed, hence a special communication and networking channel being created. One of the advantages is that these senior officials bring with them years of experience working in both public and private sectors, with in-depth knowledge of airlines operations and bureaucratic procedures. They also have extensive experience in dealing with challenging issues in both domestic and international contexts. On the one hand, personnel mobility has certainly created a tremendous amount of fluidity and informal connections between airlines and its regulators, hence, offering airlines direct access to the top policymakers in the industry (Kong, 2010). The regulator was fully exposed to the influence of the industry interest groups, which would take every opportunity to have voices heard. The corporate executives and senior managers could make phone calls any time they wanted or simply come to visit the regulator at their own discretion. On the other hand, the personnel arrangement among state-owned enterprises also sent a message to the industry not to compete fiercely between each other, as they could subsequently become adjoined to their once perceived opponents. Competition is encouraged but not to the extent where one's success is brutally achieved at the cost of the other.

Since the late 1990s, the corporatised SOE airlines have begun to develop their own strategic visions, which were quite divergent from each other. They did not necessarily share the same interest, nor pursue identical benefits. Due to historical reasons, these state-owned airlines have been handicapped from fulfilling their lobbying potential as a result of operational constraints they had inherited from the planned economy. For example, in the 1990s, Air China, China Eastern, and China Southern lobbied more for international markets entries, while the other three carriers did not show much interest in international expansion. The restricted business scope as specified in their respective AOCs, the fleet size, and the operational base had undermined their willingness to lobby the government and their capacity for lobbying.

Small, non-CAAC affiliated carriers such as Xiamen and Shenzhen Airlines did not launch any lobbying activities at all in the 1990s, due largely to their absence in the international market. Cargo carriers only started their lobby campaign in the 2000s as a result of their intention to develop international freight operations.

The continuous intimacy between airlines and the regulator did not mean that they always shared the same vision. In reality, divergent objectives and agenda appeared to be pursed since the late 1990s. In this circumstance, the regulator had to consider all arguments, make justifications, and decide on the merits of each of the arguments for a balanced policy solution. It was necessary to prioritise the country's diplomatic, military, and strategic objectives while at the same time taking into account the interest of corporate businesses.

When lobbying, individual airlines tend to force their own suggestions to be considered and accepted, which could be either identical or competing. For instance, when they all envisaged the prospective benefits and lucrativeness of new market opportunities, or competition from international rivals, their views would accord in support for free market entry or opposition to opening the market. On other occasions, they would adopt totally different views. One would seek to be awarded an international route licence and market entry entitlement, while the other would endeavour to object to any such move from their rivals.

These differences were observed before each round of bilateral discussions between China and its counterparts, when the policymaker sought for industry input to formulate its respective negotiation objectives and strategies. As discussed in Chapters 4–6, some airlines were keen to enter a specific country-pair market, while others were strongly against any new entrants, fearing that they would face more competition. When the foreign airlines voiced their intent for more access to market entry rights and capacity entitlement, the Chinese carriers were united to fight against their foreign rivals. For example, Great Wall Airline, Jade Cargo International, Shanghai Cargo Airlines, and Hainan Airlines all applied to the CAAC in 2003 and 2004 for Sino-US route licences. They cited that their move was to respond to the central government's call to promote cargo transport to enhance international trade, taking advantage of their respective business bases in Shanghai and Shenzhen, where the majority of China's export products were manufactured. In contrast, the incumbents, such as Air China, China Eastern, China Southern, and China Cargo, showed little enthusiasm about the relaxation of market-entry restrictions. In fact, they all expressed their opposition to liberalising the country-pair market, citing their primary concern being the fierce competition to be encountered from their aggressive international rivals. None of the incumbents were in favour of the increase of designation and frequencies, free commercial arrangements, or establishment of cargo hubs by the US carriers in Chinese territory. They argued that there were unused traffic rights between the two countries which needed to be exploited and that the USA's strict visa policies and application procedures imposed to Chinese citizens prevented them from generating traffic demand, hence, they could not operate as efficiently as their US counterparts.

Engagement with the government by the Chinese airport sector
Unlike the airline community which has been involved in lobbying activities in the past decades, Chinese airports have only become proactive in engaging with the government since the 21st century. This is understandable in that (1) Chinese airports were under the control of the industry regulator before 2004 which had no autonomy of its network development, and (2) Chinese airports were not managed as enterprises but as a quasi-public service organisation with no

entrepreneurial drive. Airports authorities' enthusiasm about establishing and expanding direct international links had begun to surge since 2004, when the airport reform was completed, that is, all airports, except the BCIA and those in Tibet, were handed over to local governments. The best showcase was Hangzhou airport in Zhejiang Province, which teamed up with the provincial government and convinced the industry regulator for a direct link between Hangzhou and Amsterdam. Hangzhou was about a one-hour ride by bullet train or two-hour drive south to Shanghai. Its proximity to Shanghai was seen by the Shanghai-based airlines that Hangzhou and Zhejiang province were their backyard. All international connections via Shanghai's two airports should inherently serve the catchment of Hangzhou and the province at large. The notion was challenged by the airport itself and the local government, both of which stepped up their efforts, which eventually expressed intention of KLM, the designated carrier of the Netherlands, to launch a direct service in April 2010. The route schedule was successfully included in the 2010's negotiation between China and the Netherlands.

Local government as an interest group and its influence
The local governments in China usually refer to the provincial governments and the four municipal governments in Tianjin, Beijing, Shanghai, and Chongqing, which manage the administrative regions on behalf of the SC. China's economic reform, started in 1978, decentralised its economic planning system, enabling local governments to formulate their own socio-economic development strategies that best suited their local circumstances.

Despite the central government holding a superior power over the local government constitutionally and controlling the national economic policy at the macro-level, it does not mean that local government is passive in accepting the instructions from the central government. On the contrary, the local government can exert significant influence on the central government for a preferential policy, especially in the areas of foreign investment and international trade, so as to ensure that it is empowered to achieve its economic goals. The interest of the local government is two-fold (Hampton, 2006). Economically, the local government takes sole accountability for stimulating local economic development, such as creating jobs and generating tax revenues to safeguard the well-being of the local residents. Politically, it means that the performance of the local government officials is measured against their achievements in promoting local economic growth, which directly relates to their career development. Thus, local government is so motivated and committed to their economic achievements that it would do whatever it can to achieve its goals.

When the Special Economic Zones (SEZ) were established in the 1980s in those cities such as Xiamen, Beihai, Zhuhai, and Fuzhou along the east and south coastline, flexible policies and incentives were provided, which were believed to be attractive to foreign investors to stimulate local economic growth. Soon afterwards, the local governments recognised that the preferential policies were not adequate in attracting foreign investment due to the lack of effective and convenient transportation and telecommunication systems in these cities, where the landscape is more complicated for railways, the most popular and main transport mode in China. To overcome this obstacle, local governments as exemplified by Xiamen were convinced that air links were essential for stimulating its economic growth. Thus, improving infrastructure, in particular, the air transport system, became a primary task, resulting in the local governments being inspired to finance their own airport construction and expansion and set up their own airlines. These coastal provinces, with a heavy proportion of foreign trade and a high trade-dependence ratio, not only have a significant interest in the direction of Chinese economic policy, but also have access to internal channels to exert influence on the policymaking process (Chen, 2008).

To attend to the air transport system, the local governments needed to lobby the industry regulator, that is, CAAC, which has the power to determine the construction of airports, procurement of aircraft, and issuance of AOCs and route licence. The CAAC thus became the target of the local governments, which were so determined to develop and enhance their air transport provision. The governmental officials either paid personal visits to the CAAC directly or wrote or telephoned the CAAC with their specific requests. Before the 1990s, the lobbying from the local government had mainly focused on airport construction and airlines AOCs. For example, Xiamen Airport was approved for construction in 1982 jointly financed by the CAAC and the local government. Following that, Xiamen Airlines was agreed to be set up in 1985 funded jointly by the CAAC and the local government. Shanghai Airline was set up in 1986 and Putong airport was constructed in 1998. The construction of Shenzhen Airport was started in 1990, and the Airline was set up in 1993.

In the 21st century, the lobbying turned its focus to securing international traffic rights for more direct international links. The airport localisation programme, which started in 2002 and was completed in 2004, transferred the ownership and control of all airports (except Beijing and those in Tibet) to the respective local governments (CAAC 2004), fuelled them with significant enthusiasm to seek direct international air links. Senior officials including governors would lead teams in person to meet the CAAC administrators on various occasions whenever possible to present the business cases in an attempt to convince the CAAC that their respective province should have a direct international link. Junior officials would work directly on relevant departments of the CAAC. Chinese media reported that tens of MOUs have been signed between the CAAC and the local governments, including Hainan, Jiangsu, Zhejiang, Fujian, Heilongjiang, Jiangxi, and Henan since 2003 (CAAC, 2003), whereby the CAAC pledged to give full support in their pursuit of international services by both Chinese and international carriers. The lobbying proved to be successful in many ways as almost all the major airports in these provinces became designated international airports as stipulated in the arrangements concluded between China and its counterparts such as the USA and the Netherlands.

Another showcase was Hainan Island. Hainan was separated from Guangdong province and upgraded to a provincial administration in 1988, as it was permitted some special policies to accelerate economic growth. As an island, Hainan relies heavily on air transport for external connection. Without much manufacturing or agriculture, the travel and tourism industry is the key to Hainan's economic growth. Hence establishing international links became the top priority for the local government. The strategic move of opening more international routes was initiated by Hainan Airlines in late 2002 and early 2003, which after ten years of operation, found itself in a desperate need to expand into a bigger domestic and international market. The suggestion gained immediate support from the then Party Secretary of the Province Mr. Wang Qishan, currently the country's Vice President. "A proposal to allow third, fourth, and fifth freedom traffic right operation for Hainan Province on a pilot basis" was submitted to the CAAC in February 2003, which was approved the following month.

To implement the initiative, a special "Flight Management Office" was set up. Generous incentives were offered with the aim of attracting international air operations. These incentives included operational tax refunds for the first three years of operations commencing in January 2004 and favourable tax arrangements for warehousing, logistics, and terminal operations. The local government also made ten million Chinese yuan available ($1.40 million US dollars) per annum to subsidise any airlines flying to or via Hainan, should their operations be loss-making (Hainan Local Government, 2004). In June 2004, Hainan, as a special zone in China, was documented in 2004's Sino-US Protocol, allowing operations of carriers from both countries with full third, fourth, and fifth freedom traffic rights.

To sum up, local governments in China have played a critical part in China's international trade policymaking since 1978. Although China has a unitary administrative system with the central government taking the dominant role in governing the country, local governments have enjoyed to a great extent autonomous authority in developing their own economic objectives and initiating their own trade programmes. Unlike other industries such as textiles, agriculture, and automobiles whereby the local governments have the authority to determine their operations, international air transport is centrally controlled by the CAAC, which is managed at the national level. As a consequence, there is no formal mechanism for local governments to be involved in the CAAC's air transport policymaking (Feng, 2006). In this circumstance, the influence from local governments can only be exercised when the local governments approach the CAAC directly, either in person or via other methods such as lodging a report. Local governments are able to take advantage of the central government's policy initiatives and to exert some pressure on the CAAC for a preferential international air transport policy. When they lined up the support from those airports within their jurisdiction, their strategy of joint lobbying tended to work effectively. In the 1980s and 1990s, they lobbied for approving the construction of airports, while in the 21st century, they collaborated to win international links.

Before each round of bilateral negotiations, as the preceding chapters discussed, local governments could work out the ways of securing commitment from the CAAC with their cities/destinations being included in the bilateral arrangements, regardless of the consequence of implementation. The success of lobbying, though, was subject to other factors, which included but were not limited to: (a) the availability of other policies available to support the initiative, (b) their lobbying channels and effectiveness of communication, (c) the support of their local aviation industries, and (d) their capabilities of presenting the business case.

Role of media and public opinions and their impact

The media that reflects public opinions has a significant impact on government trade policies and international negotiations (Swinnen and McCluskey, 2006). As mass media becomes the key channel of information in our society (Swinnen and McCluskey, 2006), where most people obtain information and express their opinions, it can affect the political agenda by influencing policymakers' perceptions and attitudes, thus creating a "climate of opinion" by determining what issues they consider important and how they think about these issues.

However, the ability of the media to influence trade policy was circumscribed by three conditions, as Dennis Harter, when being interviewed by O'Heffernan (1991) pointed out, which were:

a   Media's influenceability varies with the nature of the issue, with global, multilateral issues being more sceptical to media influence than bilateral or military issues;
b   Media's influenceability derives to some extent from the media's ability to stimulate domestic political forces to support or object to a policy initiative;
c   Media's influenceability varies with the prevailing political environment, although it can influence or create that environment.

Studies about the role of Chinese media in influencing the formation of China's policy towards the USA and the US-led war in Iraq revealed that the different contents covered in different media had a significant impact on the opinions formed by the Chinese public, although it had not necessarily changed China's policy output (Yu, 2005). As a mounting number of Chinese residents gain access to not only the domestic media but also global media via the internet (Lampton, 2001), Chinese policymakers are beginning to attend to the public opinions expressed and sensitive to the views presented in both domestic and international media.

There are only a handful of industry journals covering air transport, which include *China Civil Aviation*, a monthly magazine controlled by a publishing house affiliated to the CAAC and focuses on professional and technical analysis and comments on aviation issues. *International Air Transport* is another monthly magazine published by the CAAC Management Institute, which is also affiliated to the CAAC, with special interest in international aviation news. *CAAC Journal* is the only industry newspaper, controlled and published by the CAAC, four times a week, with current industry news, reports, analysis, and commentaries. It serves as the official media organ of the regulator and usually is the first to disseminate policies, reports, speeches, and remarks that the CAAC would like to disseminate. Other mainstream media publications such as *People's Daily* and *China Daily* (China's sole official English newspaper) do not cover information regarding air transport on a daily basis but do carry stories that they believe bear significance, such as aircraft ordering, air accidents, big airport construction projects, airline mergers and acquisitions, and traffic growth in China and the world. Some specialised newspapers do carry news and analysis about air transport but are totally subject to their editors' choice as to what to write and how much to write about. For example, *Caijing* (finance and economy), a monthly journal in Chinese, analysed the prospect of low-cost carriers in China in two issues in 2005 and airline mergers in 2006, respectively.

A popular private online publication focusing on air transport activities (www.carnoc.com) was established and based in Hefei, Anhui Province. Initially launched in October 1999, it gathers real-time news and reports from all sources available and is regarded as a comprehensive source of information, with an average monthly visits exceeding 4.1 million up to 2008 based on google analytics dashboard statistics. Without any public funding or bureaucratic background, the website carries objective, sometimes critical, analysis and commentaries that are not available on other state-owned media outlets, especially with respect to a particular policy matter or an event such as new route licence procedures, low-cost carriers in China, flight delays, air fare competition, as well as China's "open skies" policies.

When the 2004 Sino-US Protocol was concluded, the website was loaded with reviews and commentaries either contributed by its own reporters or other media, all of which was complaints and critiques that the Chinese Government was too hasty in opening the China–USA market. The argument was that the Chinese carriers were not prepared to face the competition from the most competitive airlines of the world. The contributors further contended that the Chinese Government ignored the concerns raised by domestic carriers, nor took any measures to protect their interest effectively (CARNOC, 2004). Interesting, though, there was little coverage on China's negotiation with the USA in 2007, China's consultation with the Netherlands, nor anything on the country's arrangements with the UK.

International industry magazines such as *Orient Aviation, Payload, Air Transport World*, and *Airline Business*, to some extent, do report on air transport activities in China, although not necessarily on a regularly basis. International newspapers that often carry news and commentaries regarding China's air transport industry include *Asian Time* and *The Strait Times* based in Singapore, and *The South Morning Post* based in Hong Kong. They tend to selectively focus on a specific event or an issue with their analysis and commentary from an international perspective.

These media publications, due to the communication language being English, does not carry the same weight as those in Chinese, the mother tongue of the people of China, nor they are considered of the same significance as the Chinese ones, which are testimonials of Chinese citizens. Nevertheless, this does not mean that the views expressed in international media are not attended, received, or appreciated. On the contrary, when a view as opposed to that in the domestic media is expressed, Chinese policy-makers tend to pay special attention to and keep a close eye on it. For example, media clippings from *The Strait Times* and *Wall Street Journal* were collected after the conclusion of the 2004 agreement to learn international reaction, which could be used for reference for future discussions.

Although a handful of industry newspapers and journals in both Chinese and English are available to report on China's air transport industry, it is still uncommon to locate any news release or reports with respect to China's international air transport policy matters. Bilateral negotiations and the agreements reached are occasionally covered. Such information is seldom found in mainstream media such as *People's Daily* and *China Daily*. One of the considerations is that any news relating to international affairs and bilateral relations is classified, which should not be released to the public. The aviation policymaker and regulator follows this practice. Another consideration is that international air transport policy is not a matter that the media can create an image of that tends to resonate with the general public, who will be intrigued to be inquisitive or allured to participate in a hot debate, thus leading to an increase in readership. International air transport policy is something that is intangible for the Chinese public, who are more concerned about flight safety and service quality, such as flight punctuality and meals provided onboard.

To summarise, although China is becoming a tourist source country, international travel propensity is relatively low. With respect to China's international air transport relationships with its counterparts, while industry professionals appreciate the significance and impact that such an arrangement would have on the overall industry and the society at large, the public is yet to recognise the benefits of liberalised arrangements. For them, a cheaper fare, regardless of the carrier, is good enough as long as they are taken to their overseas destination. For Chinese travellers to the USA, Europe, and beyond, they actually find foreign carriers cheaper and more convenient who can arrange their transit or transfer connections outside China with their extensive networks. As a consequence, limited coverage with commentaries would not bother policymakers, who would continue with their own policy agenda, nor reconsider their policy direction or priority. The media itself failed to recognise the significance of international air transport policy and its impact on the nation, the people and society at large, hence is not convinced of the value of carrying international air transport policy-related matters.

For the five country-pair cases discussed, media and public opinion apparently has not carried much weight in influencing the policy agenda nor considerations of the policy formulation. Although media coverage attracted some special attention from policymakers after the 2004 Protocol between China and the USA, it has not been influential to the extent of exerting pressure on policymakers for a policy adjustment or policy change. Therefore, it can be concluded that the role of the media and public opinion is limited in terms of influencing China's international air transport policymaking process.

Factors at the institutional level and their impact on policymaking

Scholars such as Krasner (1984) and Bailey et al. (1997) conceived that institutions have a direct or indirect impact on a nation's political process over time. Government institutions are not merely executive organs implementing the orders received from their superiors but rather play critical roles in the policymaking process which could either facilitate or distort policy objectives. They argued that institutional changes, for example, could influence the dynamics within a nation's bureaucratic apparatus, which would ultimately have an impact on the agency's agenda. Such changes would in turn affect the bureaucrats in their delivery of the policy outcome and their capabilities in achieving the policy objectives.

Although China's bureaucratic structure is different from that of other western democratic countries, its regime and system has empowered its institutions and bureaucrats with a fairly significant authority in manipulating the policymaking process, thus exerting a larger impact on the policy output. The following part of the chapter turns to the institutions involved in China's international air transport policymaking and seeks to analyse how these institutions concerned in the process relate to and interact with each other leading to a policy outcome per se.

China's aviation policy-making authorities

As a socialist country, state socialism is a politically monopolised system in which all power is held at the top and the lower levels are subordinated and dependent, deriving power as it is distributed from a higher level within the hierarchy (Huang, 2002). Neither society nor the economy retains independence from the control of the party, which has the overarching authority above the other layers of bureaucracy. The most powerful policy-guiding unit within the Chinese Communist Party (CCP) is the Political Bureau (Poliburo), and within the Poliburo, its standing committee.

As the world's largest developing country, although China has recently been recognised for its market economy status, the central government still plays a key role in intervening in economic activities by exercising autonomy from the top level. The SC is the highest executive organisation of state power and state administration. Being responsible for developing national-level strategic policies, the SC is concerned about China's international relations as well as domestic policies, economic, financial, educational, and cultural issues of strategic significance. Headed by the premier, who, together with four vice premiers and five state councillors, each of whom has their own portfolio of responsibilities for specific industries, the SC administers China's bureaucracy, including all ministries, special commissions, bureaus, and administrations affiliated to the ministries.

Regarding international air transport policy-making, the following organisations are involved, although to a different extent: National Development and Reform Commission (NDRC), MoFA, and CAAC (Figure 9.2). NDRC oversees China's overall economic policymaking from both macro and micro perspectives. The annual and five-year strategic plans for the national economy and social development spell out the goals and objectives to be achieved for industries, including air transport. The commission also monitors the implementation of the above strategic plans.

MoFA is responsible for China's overall policy concerning foreign affairs, in particular, developing and implementing the country's diplomatic policies in the international arena, although technically it reports directly to the Standing Committee of the Poliburo (Lu, 1997). The Policy Planning Department provides advice on China's international economic and financial policy based on their analysis and assessment of situations worldwide, while the Department of Treaty and Law is involved in bilateral air services negotiations, giving full support to the CAAC. Other departments overseeing a particular region such as the Department of North American and Oceania, the Department of Western Europe might be consulted internally for policy advice when a specific discussion is concerned with a specific country.

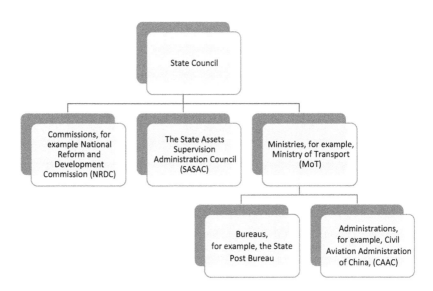

FIGURE 9.2 Relationship between Chinese bureaucracies with respect to international air transport.

Source: Compiled by the authors.

CAAC has the responsibility for developing international air transport policy directions and negotiating the output with its counterparts. The Department of Policy and Law is mainly responsible for policy research and formulation while the Department of International Affairs and Cooperation conducts bilateral air services negotiations on behalf of the Chinese government. Other government authorities, including but are not limited MoC and the CNTA, are consulted, respectively, for updated information regarding international trade and tourism so as to ensure that the principles governing the negotiations are adhered to, the bottom line is upheld, the national and public interest is considered and protected, and the negotiation outcome aligns with the nation's overall diplomatic and strategic objectives.

Structural changes

Parallel to its economic reform, China's political reform also follows an evolutionary approach, which has seen the downsizing and streamlining of China's bureaucracies over the past three decades. CAAC experienced several structural changes, seeing itself separated from the Air Force in the early 1980s and merged with the MoT in 2008. Regardless, CAAC's core responsibilities, that is, regulating the domestic and international air transport operations from economic and safety perspectives remain unchanged.

Until 1978, CAAC was part of China's Air Force, overseeing non-military air transport activities. As a quasi-military administrative agency, CAAC, receiving instructions from the Air Force, which controlled its human resources, finance, and airspace, regulated and executed civilian and commercial air transport operations in both domestic and international markets.

First structural change in the 1980s

The first structural change came in 1980 when Deng Xiaoping called for civil aviation to be corporatised, resulting in the separation of CAAC from the Air Force. It became a sub-ministry level government agency to continue with its responsibilities of regulating and operating the air transport industry. Its reporting line was switched from the Air Force to the SC. The intention was to help CAAC establish a concept of modern management of the airlines industry to demolish the monopoly control over the industry. This was not an easy task to accomplish due to the fact that CAAC had been traditionally overshadowed by the military that it required Deng Xiaoping's personal efforts and intervention (Williams, 2009).

The launch of the six major airlines[2] had not been achieved until 1988, almost eight years after Deng's call to reform the civil aviation sector. CAAC retained its authority to regulate economic activities of the sector such as market entry, capacity, and pricing on both domestic and international operations. It also determined the fleet size for each individual airline, and funding for airport construction and expansion and air traffic management facilities. It negotiated bilateral air services agreements and any amendments thereto, and oversaw the implementation of such arrangements. For the Sino-US and Sino-UK markets, CAAC was replaced by Air China in 1987, with China Eastern becoming the second designated carrier in the Sino-US market.

Structural changes in 1993

In 1993 CAAC experienced a radical change: it was upgraded to a ministry-level government agency under the direct control of the SC. The justification was that the rapid growth of the industry over the past decade caused severe safety concerns,[3] which alarmed the central government forcing preventative measures to restrain the overenthusiasm towards air transport. One of the measures taken was to strengthen the position of the CAAC within the national bureaucratic system to bring it into the mainstream of the agencies of the SC, with Mr Chen Guangyi (December 1993 to June 1998), the then Party Secretary of Fujian Province, being appointed as the first ministerial-level administrator. The move intended to empower CAAC with a more powerful regulatory authority over the industry.

The early years of Chen's tenure failed to dampen the nation's enthusiasm for air transport. In 1995, there were 43 airlines operating in China, most of which were established by the local provincial and municipal governments, with some smaller ones operating with only two to three Russian-made airplanes. Price wars started with airlines selling tickets at less than 50 percent of the published fares, resulting in the CAAC, together with the National Planning Commission (NPC) being responsible for the price reform of all industries in China, to promulgate a series of regulations in an attempt to control air fares. The three big carriers—Air China, China Eastern, and China Southern—at the same time consolidated their market power carrying 56 percent of the total paying traffic in 1997, and providing 79 percent of the domestic services.

Dramatic downsizing in 1998

Failing to regulate the industry effectively, CAAC was further challenged by another round of administrative reform in 1998, when the agency was required by the SC to halve its size, cutting its staff numbers from 450 to 252 without compromising on its responsibilities. For the first time in CAAC's history, the bureaucrats, a good percentage of whom were ex-military officers after years of civil service in the government agencies, were concerned about their personal career prospects, panicking with a desperate desire to know where they would go to work the following day. The functional departments, being involved in the economic regulation of both domestic and international operations, were unable to retain all of their veteran staff due to the quota requirements. To cope with the workload, CAAC had to make arrangements with airlines and other CAAC-affiliated organisations, which were able to provide temporary human resources without affecting the employment relationship. The reform was not completed until 2002, when CAAC was deprived of some of its major responsibilities. For example, SASAC rather than CAAC would appoint top management for the three big carriers that were cherrypicked to merge out of the previous six trunk airlines. CAAC was also deprived of the responsibilities of managing all national airports financially and operationally (except Beijing Airport and those in Tibet).

Between 1998 and 2007, CAAC was led by two ministers, with Liu Jianfeng serving from June 1998 to May 2002 and Yang Yuanyuan succeeding Liu until December 2007. With a background as a pilot, Yang was seen as open-minded, fully embraced with the trend of international air transport. This was in contrast to his predecessors Chen and Liu. He was well received in the international aviation community, being nicknamed as triple Y, thanks to his effective communication skills and straightforward personal style. Being hailed as "the one" by IATA's then CEO (IATA, 2005), indicated that Yang would be the leader to take China's aviation into the next era, embracing a more liberalised approach to the sector.

During Yang's period, cargo was proactively and promoted by CAAC to support the country's export-oriented economic growth. Domestic private capital was also allowed into the airline industry, bringing about several start-ups which adopted the LCC business model. Industry consolidation resulted in three megacarriers dominating China's domestic and international operations with an aggregate 80 percent market share (Williams, 2009). By joining international alliances including Star Alliance and SkyTeam, the three trunk airlines were ambitious in developing organisational capabilities to compete on a global scale.

The move to bring CAAC under the MoT

In 2008, the SC decided to further advance its political reform by announcing to integrate CAAC into the MoT, thus downgrading its status to a sub-ministry level, with its scope of responsibilities being unaltered. The intention was to streamline the regulatory functions to minimise the overlapping. The new MoT would oversee all modes of transportation including road, water, air, and postal services, except railway. Administrator Li Jiaxiang was appointed vice minister of the MoT whose portfolio included air transport. With a military background, Li was first involved in civilian air transport in November 2000 when he was appointed party secretary of Air China Co Ltd. In two years' time, Li was promoted to party secretary and vice president of Air China Group, one of the three mega-carriers which took over Southwest and Zhejiang Airlines. In December 2007, Li replaced Yang Yuanyuan to become the 12th administrator of CAAC. Believing that competitiveness of Chinese airlines will only be gained and developed through further industry consolidation such as merger and acquisition, he advocated cooperation, integration, and alliance among carriers to forge two or three large Chinese carriers with international competitiveness (China Daily, 2008).

The downgrading required CAAC to report to the MoT rather than directly to the SC, indicating that there would be a different administrative procedure for CAAC to follow and a longer lead time for CAAC to receive any feedback or instructions to take further action. In China, bureaucratic ranks are so important that negotiations and coordination can only be achieved among bureaucrats who consider themselves equal and who have the authority and power to instruct and order their subordinates (Kong, 2010). It is anticipated that the relationship between the MoT and CAAC would replicate to some extent the management system of the USA, as found in the DoT and FAA (Williams, 2009). On operating within the new framework while trying to integrate itself into a new organisation, the CAAC was not assured of its future status, with speculation that the Ministry of Railways and the current MoT will eventually merge into one super-agency, looking after all modes of transportation for a better coordination and planning of the system. With this speculation down the line, it is hoped that a concerted culture can be developed so as to support the organisational change without affecting the performance of the staff.

Impact of structural changes on the country's international air transport policymaking

Bureaucracies are created by governments to implement policy. However, they have a significant role to play in the process of policymaking which cannot be entirely separated from policy implementation. Organisational theories have argued that bureaucratic culture, once established, is difficult to change even when underlying social forces continue to evolve, as particular institutional arrangements create privileged positions for individuals and groups who work to perpetuate those arrangements (Ikenberry, 1988). Individuals within the organisations would seek to preserve and protect their missions and responsibilities even when the specific circumstances that brought the organisation into existence have changed (Ikenberry, 1988). The cost and uncertainty involved in the organisational change usually generate countervailing incentives for the maintenance of the existing organisations (Ikenberry, 1988). Bureaucracies with their featured culture, rituals, and routine responsibilities have a direct impact on the way the individuals involved perceive interest and value, both domestic and international (Ikenberry, 1988). Such organisational characteristics can bias the way in which the imperatives of an international system are perceived and acted upon, hence, influencing the capabilities of officials to diagnose the environment to formulate the policy (Ikenberry, 1988). Bureaucratic organisations with their own power and resources could stir up the policy process by shaping the perceptions of political actors, thus affecting the policy outcome.

Unlike other ministries or governmental agencies which have been abolished, reshuffled, merged, created, or restructured, with their functions and roles being deprived, altered, and adjusted through rounds of administrative reforms, CAAC has survived various rounds of structural changes and kept hold of its core portfolios of duties and responsibilities as an economic and safety regulator of the air transport sector. The constant but not consistent restructuring resulted in it being upgraded and downgraded with the number of employees being increased and decreased. While it has to spin off its non-governmental obligations to focus on the development of its regulatory capabilities, the restructuring has dramatically weakened its administrative, human resource, and financial capacities. It is no longer the authority to dictate management and operational issues of the industry; to appoint/relocate/second top-level management to airlines and airports; or to collect the proportion of operational profit from the industry.

The deprivation of these powers has had a effect on the competencies and negotiation capabilities of the institution. The motivation, morality, and performance of the senior officials as well as staff were affected. The inconsistency of policies and ever-changing procedures also affected its implementation strategies and outcome. Over the years, CAAC has been able to retain some key officials who have extensive industry and international experience and are given the leading role in the policymaking process. However, the regulator also experienced loss of veteran staff, who were replaced with green hands, who needed time to develop expertise. While it is too soon to justify how this has affected policymaking, it is widely accepted that more experienced policymakers are more capable of managing the situation to achieve optimal policy outcomes.

CAAC developed its regulatory functions out of a quasi-military background. The learning curve for CAAC to transform from a combined regulator and operator in the 1980s to an effective regulator in the 21st century is long and deep. In this process, the regulator has been the driving force for aviation growth, while developing its own organisational capabilities. Unlike other industries such as petroleum and telecommunication, whereby the regulatory reform has deprived the key regulatory functions of the bureaucracies, CAAC has managed to uphold its entire economic control and safety management over air transport activities. While it is required to seek instructions and advice from other ministries, it only involves a handful of such government agencies as the SC and MoFA (see Figure 9.3). Such a structural simplicity has enabled CAAC to devote majority of its resources to study on policy matters rather than coordinating for a consensus. It has also allowed CAAC to work on its own policy initiatives when negotiating with its foreign counterparts without much external resistance, which could be out of its control. CAAC's primary consideration is whether the liberalisation arrangements and the exchanges of traffic rights are in alignment with the nation's overall diplomatic and strategic policy objectives.

Summary

Chapters 7, 8 and 9 have identified the factors at different levels and analysed how these factors converged on and interacted with each other to affect China's international air transport policymaking. It started with identifying the factors at

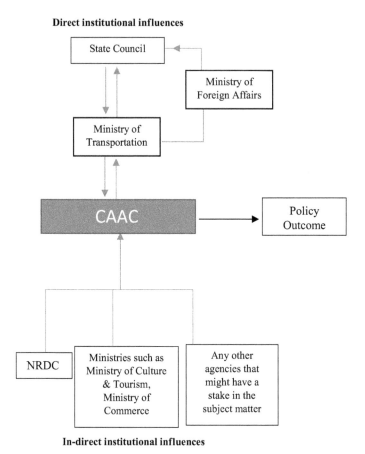

Direct institutional influences

In-direct institutional influences

FIGURE 9.3 Decision-making reporting line and CAAC's institutional influences.
Source: Compiled by the authors.

international and regional levels, followed by establishing factors at domestic and institutional levels. We argue that many factors at various levels have played a role in influencing the country's international air transport policymaking outcome to a different degree, during the policymaking process, as reflected and executed in individual country-pair bilateral air services arrangement negotiations. However, these factors are constrained by time and circumstances, with some factors playing a bigger role at certain times but not all the time.

The international air transport policy, an element of China's national economic policy, exemplifies the nation's changing view towards liberalisation of international trade. Over the years, China has experienced a dramatic change in terms of its social and economic system, with its policy direction transforming from being protective and conservative to being pro-competitive and advocating liberalisation. The policymaking process is also becoming more open, transparent, and pluralistic, subject to the influence of many variables at different time and levels.

Notes

1 There were 196 big SOEs under the control of SASAC in 2003 when the authority was first formulated with total assets of seven trillion Chinese yuan. In 2009, the number of SOEs decreased to 123, but the total assets increased to 21 trillion Chinese yuan; 30 of them were among the world's top 500 companies.
2 The six major airlines were Air China established in Beijing in July 1988, China Eastern in Shanghai in June 1988, China Northwest in Xi'an in December 1989, China Northeast in Shenyang in June 1990, China Southwest in Chengdu in December 1987, and China Southern in Guangzhou in February 1991.
3 Between July 1992 and December 1993, nine aircraft crashed as a result of industry expansion causing a serious shortage of professionally qualified and experienced pilots.

References

Aggarwal V. K., and Lee S., (2011), The domestic political economy of preferential trade agreements in the Asia-Pacific, in: Aggarwal V. K., (edited), *Trade Policy in the Asia Pacific, the role of ideas, interests, and domestic institutions*, (2011), Springer, New York and London, pp. 29–48.

Bailey M., Goldstein J., and Weingast B. R., (1997), The institutional roots of American trade policy: politics, coalitions, and international trade. *World Politics*, 49(3), 309–338.

CAAC, (2003), *Statistical Data on Civil Aviation of China, 2003*, Civil Aviation Press of China, Beijing.

CAAC, (2004), *Statistical Data on Civil Aviation of China, 2004*, Civil Aviation Press of China, Beijing.

CAAC, (2007), CAAC's policy proposal with respect to the implementation of the Party's Principles of the Sixth Plenary Session of Sixteenth Congress, July 2007, available at www.caac.gov.cn/B7/200707/t20070719_6569.html

Cantle K., (2010), Government to subsidize CA's Air China Cargo Acquisition, 20th January, 2010, available at www.atwonline.com

CARNOC, (2004), Civil Aviation Resource Net of China, available at www.carnoc.com/

Chen Z., (2008), Coastal provinces and China's foreign policymaking, available at www. irchina.org/en/pdf/czm5.pdf

China Daily, (2008), Air China parent seeks partnership, not merger with China Eastern, available at www.chinadaily.com.cn/china/2008-01/10/content_6385332.htm

China Labour Bulletin, (2007), Reform of state owned enterprises in China, available at https://clb.org.hk/en/content/reform-state-owned-enterprises-china

Cliff R., Chad J. R.O., and David Y., (2011), *Ready for takeoff: China's advancing aerospace industry*, RAND Corporation, Santa Monica, CA.

CPEC, (2017), New Gwadar International Airport., available at www.cpec.gov.pk

Ernst D., and Naughton B., (2008), China's emerging industrial economy, in: McNally C. A., (edited), *China's emergent political economy, capitalism in the dragon lair*, (pp. 39–59), Routledge, London and New York.

Ernst & Young, (2016), China go abroad: key connectivity improvements along the Belt and Road in telecommunications & aviation sectors, EY, Shanghai. available at http://english.peopledaily.com.cn/200506/27/eng20050627_192495.html

Feng H., (2006), *The politics of China's accession to the World Trade Organization*, Routledge, London and New York.

Gao Y., and Tian Z., (2006), How firms influence the government policy decision-making in China. *Singapore Management Review*, 28(1), 73–85.

Hainan Local Government., (2004), Business, available at: http://en.hainan.gov.cn/eng-lishgov/Business/index_13.html

Hampton A., (2006), Local government and investment promotion in China, Proceedings at "Public action and private investion" of the "centre of the future state", hosted by the Institute of Development Studies, University of Sussex, December 2006.

Han A., (2008), Building a harmonious society and achieving individual harmony. *Journal of Chinese Political Science*, 13(2), 2008, available at www.springerlink.com/content/r3jk257016v23911/fulltext.pdf

Hao Y., (2005), Introduction: influence of social factors: a case of China's American policy making, in: Hao Y., and Su L., (editors), *China's foreign policy making, societal force and Chinese American policy*, (pp. 1–18), Ashgate Publishing Limited, Aldershot.

Hao Y., and Su, L., (2005), *China's foreign policy making, societal force and Chinese American policy*, Ashgate Publishing Limited, Aldershot.

Huang, Y., (2002), Managing Chinese bureaucrats: An institutional economics perspective, *Political Studies*, 50, 61–79.

IATA, (2005), IATA honours Chinese Minister with gala award, IATA press releases, on 30th May 2005, available at www.iata.org/pressroom/pr/Pages/2005-05-30-03.aspx

Ikenberry G. J., (1988), Conclusion: an institutional approach to American foreign economic policy, in: Ikenberry G. J., Lake, D. A., and Mastanduno, M., (editors), *The state and American foreign economic policy*, (p. 219), Cornell University Press, Ithaca, NY.

Ji X., (2010), The analysis and understanding of the discussion about "state-owned progress and private retreat" in its right perspective, April 19th, available at www.sasac.gov.cn

Kennedy S., (2005), *The business of lobbying in China*, Harvard University Press, Cambridge, MA and London.

Kong B., (2010), *China's international petroleum policy*, Praeger Security International, California and Oxford.

Krasner S. D., (1984), Approached to the state: Alternative conceptions and historical dynamics. *Comparative Politics*, 16, 223–246.

Ku S. C. Y., (2006), China's changing political economy with Malaysia: a regional perspective, in: Yeoh E. K., and Chung H. K. (editors), *China and Malaysia in a globalizing world, Bilateral relations, regional imperatives and domestic challenges*, (p. 32), Institute of China Studies, University of Malaysia, Kuala Lumpur.

Lai H., (2010), *The domestic sources of China's foreign policy, regimes, leadership, priorities and process*, Routledge, Abingdon.

Lampton D. M., (2001), *Same bed different dreams, managing US-China relations 1989–2000*, University of California Press Ltd., Berkeley, Los Angeles and London.

Le T., (1997), Reforming China's airline industry: from state-owned monopoly to market dynamics. *Transportation Journal*, 37(2), 22 December, 45–62.

Li J. M., (2008), Interview conducted in Beijing June 2008.

Li J. M., (2009), Interview conducted in Beijing June 2009.

Lu N., (1997), *The dynamics of foreign-policy decision making in China*, Westview Press, Boulder, CO and Oxford.

Meyer M. W., (2008), China's second economic transition: building national market. *Management and Organisational Review*, 4(1), 3–13, Oxford: Blackwell.

O'Heffernan P., (1991), *Insider perspectives on global journalism and the foreign policy process: Mass media and American foreign policy*, Ablex Publishing, Westport, CT.

Park S., and Luo Y., (2001), Guanxi and organisational dynamics: organisational networking in Chinese firms. *Strategic Management Journal*, 22, 455–477.

Pekkanen S. M., and Tsai K. S. (2005), Late liberalisers, comparative perspectives on Japan and China, in: Pekkanen S. M., and Tsai K. S., (2005) (editors), *Japan and China in the world political economy*, Routledge, Abingdon and New York.

People's Daily, (2005), Building harmonious society crucial for China's progress: Hu, 27 June, 2005, available at Li X., and., Brødsgaard K. E., (2013), SOE reform in China: past, present and future. *The Copenhagen Journal of Asian Studies*, 31(2), 54–78.

State Council, (2017), 74 new airports to be completed by 2020, available at http://english. www.gov.cn/state_council/ministries/2017/02/19/content_281475571877834.htm

Swinnen J. F. M., and McCluskey, (2006), *Trade, globalisation and the media: introduction*, Blackwell Publishing Ltd., Journal compilation, Oxford.

Tsai K. S., and Cook S., (2005), Development dilemma in China: socialist transition and late liberalisation, in: Pekkanen S. M., and Tsai K. S., (editors), *Japan and China in the world political economy*, Routledge, Abingdon and New York.

Williams A., (2009), *Contemporary issues shaping China's civil aviation policy, balancing international with domestic priorities*, Ashgate Publishing Ltd, Farnham.

World Bank, (2012). *China 2030*, The World Bank, Washington, DC.

XiamenAir., (1996), About uS: 1996, available at www.xiamenair.com/en-us/aboutUs/annualEvents.html

Xinhua, (2018), China's aviation industry contributes to connectivity along Belt and Road, available at www.xinhuanet.com/english/2018-08/10/c_137381825.htm

Yu Y., (2005), The role of the media: a case study of China's media coverage of the US war in Irag, in: Hao Y., and Su L., (editors), *China's foreign policy making, societal force and Chinese American policy*, Ashgate Publishing Limited, Aldershot and Burlington, VT.

10

CONCLUSION AND CONTRIBUTION TO KNOWLEDGE

Conclusion

Foreign trade exists in two dimensions. Outwardly, it consists of economic transactions in which goods and services are exchanged for money by persons or entities in two countries. Inwardly, it is a political process in which difficult choices must be made between competing values, priorities, and objectives (Cohen et al., 2003). Trade policymaking, thus, is a concerted effort to reconcile and balance a nation's economic goals and political ambitions. The interplay between these governmental decisions is critical to addressing its domestic and foreign concerns. A state government needs to balance conflicting intellectual viewpoints, trade off potentially diverging domestic interests with foreign demand, and honour international rules and regulations. International air transport policymaking is poignantly a testimony to these dynamics.

International air transport has been especially politically volatile and always attracts public interest, discussion, and controversy (Snedden, 1981). It has remained as one of the most debated topics in governments, attracting attention beyond its economic significance. For many politicians, air transport has become a battleground for conflicting political philosophies (Snedden, 1981). To ensure what has been achieved constitutes a "good international air transport policy," it requires the government to sort through considerations of the best national interest, conflicting intellectual viewpoints, compatible or competing domestic interests, and international influence. Sovereign governments also need to ensure that they honour international laws and rules that govern the global operations of air transport. The multifaceted nature of the subject matter makes the policymaking of international air transport a more sophisticated political process with significant economic substance.

There are a wealth of studies on international air transport, with the majority focusing on the economic impact of deregulation since 1978. Research on international air transport as an international trade in services has been more recent. In particular, studies on policymaking in the domain of international trade in services have been limited, fragmented, and unsystematic. Examining China's international air transport policymaking from an international political economic perspective is a brand new endeavour.

This research treats international air transport as an international trade in services and uniquely assesses China's international air transport policymaking from a political economic perspective. It aims to identify the factors that have had an impact on the country's policymaking process, establishing whether these factors have been evolving over the years and determining how they have been interacting with each other, leading to the policy outcome. It further intended to shed light on the country's international trade policymaking and its international behaviour. In this way, the goal was to provide testimony to illustrate how China's political and economic reform and current globalisation trend have affected the country's political system, which itself has been evolving. More than 30 years of reform have transformed China from a closed planned economy to an open market economy. Globalisation has increased the country's exposure to the international environment. In this kind of dynamic circumstance, the Chinese government has also adjusted its roles and functions to become more adaptable to meet both domestic and international challenges.

The Micro-Macro Linkage Approach was applied to analyse China's international air transport policymaking. Using several country-pair markets as case studies and through a critical analysis of data collected from secondary sources and in-depth interviews with those who have participated in the process of policymaking and/or have been personally involved in the bilateral air services agreement negotiations, this research has led to the following findings:

- The factors at all levels and from various sources, that is international, regional, bilateral, domestic, institutional, and individual, have affected the country's overall international air transport policy direction and policy goals;
- International regimes are precursors that oblige China to take actions to adjust its international air transport policy directions while domestic considerations dictate the country's overall international air transport policy goals and vision;
- Regional and bilateral relationships between China and its counterparts affect the country-pair policy objectives, while the timing and circumstances embracing each round of bilateral negotiations shape the country-pair market policy outcome;
- The industry regulator at the ministerial level, namely the CAAC in this case, is the key initiator, negotiator, and administrator in the process of policy formulation and implementation. Although it requires coordination and consultation between institutions and agencies, it is not as fragmented as the case in energy issues discussed by Lieberthal and Oksenberg in 1988.

- The state elites are not noticeably personally involved in the process of daily management although their attention might be drawn occasionally at certain points where a higher-level instruction and decision is sought;
- The factors that have an impact on the policymaking process have been evolving with new variables emerging as environmental changes. The weight of the identified factors in the policymaking process is related to and constrained by environment, time, and issues. Some factors are able to play a bigger and more decisive role than others in certain circumstances but may not play a part in other instances. No factors have remained dominant in dictating what the policymakers will come up with as a policy outcome, but rather, it is a combination of the factors that have an influence. A particular policy outcome results from the careful and prudent considerations of various factors which are traded off by policymakers.
- The international air transport policymaking process is becoming more dynamic, being more open, pluralistic, and transparent involving more stakeholders seeking a fair and balanced consideration of their views and comments.

Factors at all levels have influenced China's international air transport policymaking

The International air transport sector has become more liberalised with the majority of economic restrictions being removed through the USA's open skies initiatives. By March 2009, at the bilateral level, a total of 157 open skies agreements had been concluded involving 96 member states (ICAO, 2010). These arrangements together with regional liberalised agreements covered approximately 32 percent of country-pair markets with non-stop international passenger services and almost 57 percent of the frequencies offered (ICAO, 2010). At the national level, some 16 member states adopted open skies policies, which have unilaterally liberalised foreign airlines' market access to their territories, whether holistically or partially. Some member states have adopted more staggered approaches instead of launching review processes of their air transport policies in light of the global trend towards increased liberalisation (ICAO, 2010). By October 2012, over 400 liberalised agreements had been concluded between 145 member states (ICAO, 2013).

Although China is not one of those countries that have fully liberalised its domestic and international air transport markets, it has made significant progress in removing the economic constraints on air transport operations. It has transformed itself from a closed-door, centrally controlled economy to an open market economy. As it is integrating into the world economic system, it is also prone to dynamic factors from various sources that would influence its policymaking process and shape its policy outcome. Such dynamics not only increase the uncertainty of the policy outcomes, but also the diversity of actors involved and thus the fluidity of the negotiation processes.

International regimes are precursors that oblige China to take actions to adjust its international air transport policy directions

At the international level, intergovernmental institutions such as ICAO and the WTO are instrumental in affecting China's overall international air transport policy direction. The International political system, including international organisations, international law, informal norms, as well as non-intergovernmental organisations, is an important structure that constrains and shapes a country's foreign policy (Simmons and Martin, 2002). International considerations are prominently factored into a nation state's decision-making to the extent that domestic critics have expressed irritation and argue that a nation state's own identity is increasingly shaped by the international political context. When trying to balance the two ends of the spectrum, that is, domestic benefits and international obligations, policymakers are caught in the condition of interdependence, where they need to seek the best second to none solution.

China's entry into the WTO in 2001 has compelled the country to formulate policies and adjust its legislature to be in compliance with the world economic system. WTO membership forced the country's very top leaders, such as Jiang Zemin and Zhu Rongji, who were the nation's president and premier at the time, to take proactive measures to implement the commitment, resulting in a fundamental structural change in society, the system, and the industry. Similarly, ICAO also obliged the country to respond to its call for liberalisation. As a consequence, China took immediate action to issue government directives and regulations to remove the economic constraints on air transport operations.

Regional and bilateral relationships between China and its counterparts affect the country's country-pair policy objectives

Dealings with ASEAN and the peripheral economies in Northeast Asia have been identified as an important factor at the regional level. They are not only geographically closer to China but are politically important to the country as they are economically interdependent on each other. Liberalisation between these economies and the USA, between them and China, and between themselves have produced both a positive and negative impact on China's view towards open skies in bilateral markets, forcing the government to re-evaluate its policies towards the industry, which eventually led to the adoption of a progressive approach to embrace liberalisation. While the impact on China's overall air transport policy is yet to be determined, it certainly has influenced its policy outcome and negotiation results in the China–USA market.

At the bilateral level, political relations between China and the respective countries have affected their respective political stance, economic policies and, more explicitly, the air transport policy. Establishment of diplomatic ties led to the establishment of air links, while disagreement and divergent views on certain politically sensitive issues led to the suspension and interruption of air links. When economic relationships become closer and interdependent, demand for air transport grows rapidly. The ever-growing bilateral trade has stimulated the desire for closer air transport relations, requiring governments at both ends to respond to market forces. The trade relationship has certainly influenced the preparation, objectives, strategy, and outcome of each round of negotiations. In addition, it affected the mechanism of market liberalisation, the pace of the liberalisation on a particular bilateral market, and the approach as to how the liberalised arrangements would be introduced.

Domestic considerations dictate the country's overall international air transport policy goals and vision

At the domestic level, the central government's national policy initiatives have been identified as a decisive factor, determining the country's international air transport policy direction and objectives. This is likely a reflection of the special features of China's social and economic system in that bureaucracies and institutions are obliged to formulate industry-specific policies to mirror the strategic mission of the central government. The CAAC, as a government agency, has the autonomy to develop industry-specific policies and to ensure these policies are in alignment with the nation's strategic vision, politically and economically. The decisions about political and economic reform of industries exert a great influence on the industry regulator's strategies and capabilities of delivering the policy outcome. The regulators need to determine to what extent their policies should protect the industry and to what extent their policies should promote competition by creating a fair-playing field. The relationship between the central government and the industry regulator is an important element in understanding the impact on the industry's policy outcome.

Also at the domestic level, societal forces such as media and interest groups, which include airlines and airports, as well as local governments, have had an impact on China's international air transport policymaking, in particular, on the objectives, strategies, and results of the bilateral negotiations on country-pair markets. However, to what extent they are capable of persuading bureaucrats to act on behalf of an individual organisation's best interests is subject to many factors, including:

- Whether they have effective access to the policymakers;
- How they present their case;
- Whether their request aligns with the national strategic and economic policy direction;
- Whether their request aligns with the local economic development goal;
- The timing of their lobbying, namely when they lobby for what; and
- To whom their lobbying is directed.

One of the differences between China and other industrialised countries in the West in terms of the influence of industry interest groups is that private business plays a bigger role in western countries, while in China, it is the state-owned enterprises that are more active in their lobbying. Greater constraints have been imposed on policymakers' preferences due to the incumbent regime, that is, policymakers also need to take into account their personal political career development and tread on some fine lines carefully; thus, various views need to be taken into account before reaching a policy outcome (Lai, 2010).

It is also worthy to note that international elements have joined the force of interest groups to lobby the Chinese government in its international air transport policymaking. Such cross-border lobbying, taking different formats, has gained its momentum after China's access to the WTO as a member. These foreign elements include, but are not limited to, foreign local governments and their agencies, such as a state or a council government, the tourism board, foreign airlines, and airports. They would either approach the local Chinese government, whose provincial capital city was their target direct flight destination, or Chinese carriers whom they wanted to provide air services, or to the CAAC directly seeking to support a liberalised bilateral arrangement between the country-pair markets. They would also work together with their target Chinese partners to develop "collaborative plans," which then would be tabled by the Chinese partners during the negotiation, thus rendering some pressure on the Chinese policymakers. These kinds of indirect, round-about lobbying strategies have proved to be effective, as the Chinese government was convinced with achievable outcomes that would benefit its industry and the community.

Media with a special focus on the air transport industry has also been established as a factor, although the aviation-oriented media has not played a bigger role in shaping the public opinions that could affect the country's liberalisation policy outcome.

Bureaucracies at the institutional level are instrumental in crafting policy objectives and executing strategies to achieve policy outcomes

At the institutional level, CAAC has played a pivotal role in formulating the nation's air transport policy, being charged with responsibility by the SC for regulating the air transport sector. It determines how to articulate the nation's international air transport policy direction and objectives to both domestic and international community. Essentially, the CAAC needs to establish the most appropriate approach to align its air transport policy with the central government's strategic objectives. It needs to assess all the requests from interest groups and to determine the trade-off. It also takes sole responsibility for formulating bilateral negotiation strategies and tactics, and is accountable for the negotiation outcome, with the ultimate aim of ensuring sustainable benefits for the industry and society at large.

The several rounds of structural reform have not affected the CAAC from retaining its core responsibilities. Unlike the country's other industrial bureaucracies

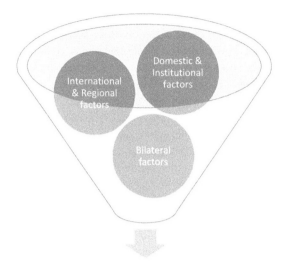

Policy outcome

FIGURE 10.1 Factors at all levels and their interactive impact on the policy outcome. *Source*: Compiled by the authors.

whose powers are fragmented and shared with one another, the CAAC has the sole responsibility for formulating the nation's air transport policymaking. It is the focal point where all factors, regardless of their levels, interact, converge, and are filtered by bureaucratic policymakers. It is here that the policy is debated, crafted, formulated, and implemented (Figure 10.1).

This centralised, tightly controlled policymaking approach is in contrast with that of China's policymaking in the petroleum and energy sector. Several government institutions have been involved in the decision-making, who had to invest significant resources in bargaining with each other to build consensus. The failure to coordinate effectively due to conflict of interest would result in a lengthy process of policymaking and the intended policy outcome tends to be compromised (Kong, 2010).

Evolution of factors and their changing nature

As the French aphorism goes, *"the more things change, the more they remain the same"*. In the natural world, change represents the universal constant. In the political context, the policy agenda of a nation state continuously expands, either in content or in scope (Cohen et al., 2003). They are refined, adjusted, adapted, and revamped to reflect the changing nature of the environment. The changes are best viewed as additive, incremental, and continuous rather than abrupt, radical, or transformational (Kong, 2010). This research has established that China's international air transport policymaking is not immune to this axiom.

International regime becomes a factor only after China becomes a signatory member state

The International regime has had an unprecedented impact on China's international air transport policy ever since the country became a member state. The legally binding rules and regulations have pivoted China in the direction of that incremental and continuous change. This is best exemplified by the country's membership of ICAO since 1972 and the WTO since November 2001. China's obligation to comply with the internationally agreed conventions have a fundamental impact on the country's political, legal, regulatory, and economic systems. While international regime is not a new factor in influencing the country's international air transport policymaking, the weighting of the role that both ICAO and WTO have played have differed substantially. When comparing these two, the WTO membership is more a changing agent that did not exist before the 21st century, which has played a bigger part in propelling China to change its approach from a protective regime to a liberalised mechanism.

Regional factors are an emerging element since China has recognised its significance

Being involved in regional economic integration activities has had a great impact on China's consideration of its political and economic policy directions, as geo-politics not only affects a nation's stability but also its economic prosperity. Participation in region-wide political and economic programmes enabled the country to gauge the external environment and attest its policy initiatives to gain experience to be applicable at the global level. This factor is of an evolving nature since its influence had not been exerted before 1997 when China was not actively involved in regional forums.

Factors at the bilateral level have been persistent but the country-pair governments and policymakers have the ability to weather through the differences to pursue optimal economic benefits

Bilateral relationships have played an irreplaceable role affecting the country's international air transport policy on the individual country-pair market, if not at the national scale. It determines the level of liberalisation on the individual country-pair market and affects China's strategy, negotiations, and outcome. Such a bilateral relationship has remained an ongoing factor ever since diplomatic ties were established between China and its counterparts. Since the beginning of the 21st century, various types of exchange programmes, such as industry training, cultural exchange, ministerial/business summits, and dialogues, were added to supplement the bilateral discourses and have become sources of influence. Different programmes targeting different levels of participants have apparently exerted different pressures on different country-pair markets.

Factors at the domestic level have been officially recognised and the lobbying activities are becoming institutionalised

Interest groups are able to have their voices heard through various channels and their preferences accepted through their persistent and timely lobbying. Their success depends on whether their preferred policy outcome aligns with the policy direction of the central government and the industry regulator. On this basis, those who used to be inactive are getting more proactive in the process of lobbying as the country's political reform is getting more intensified. The lobbying activities were first seen in the policymaking process from 1987 when the airline industry was initially restructured. Local governments have been active since the 1980s although the focus of their lobbying evolves to align with their local economic interests. Airports as interest groups had not been integrally involved until after the 21st century when they were spun off the CAAC's control.

Institutions and bureaucrats who have the power to trade off the factors and shape the policy outcome have remained pivotal players in the policymaking process

CAAC has remained the policymaker ever since its separation from the Air Force in the late 1970s, regulating the air transport sector from both economic and safety perspectives in China. The title alteration, structural reform, and administrative status adjustment in the past few decades (Figure 10.2) have not affected its portfolios of responsibilities, but rather, its power has been strengthened. This finding corresponds to Kong's study (2010) on China's policymaking in the petroleum industry, whereby he observed that it was the government agencies at the ministerial level that initiated and formulated policy directions and oversaw its implementation. The industry regulator was a key change agent in this process, enabling the structural reform taking place throughout the sector across the country.

When policymaking is made at the ministerial level, the structure of the agency, and the administrative status and experience of the bureaucrats in office exert significant impact on the policy outcome. This finding is not in conformity with what Lieberthal and Oksenberg (1988) found out with respect to the policymaking in China on energy matters. Lieberthal and Oksenberg concluded that the decision-making on energy matters in China was a fragmented process involving quite a number of government agencies, who shared the power of policymaking. These agencies invested significant resources in bargaining with each other to build consensus, thus, making the process diffused, protracted, and disjointed. To enable the policy to be passed as national policy, the bureaucrats at the ministerial level were capable of manoeuvring the situations to force an issue onto the agenda of the highest leaders at the top Politburo level.

Unlike the USA where the President and House of Senate, or the EU where the Council of Ministers would, from time to time, attend to international air

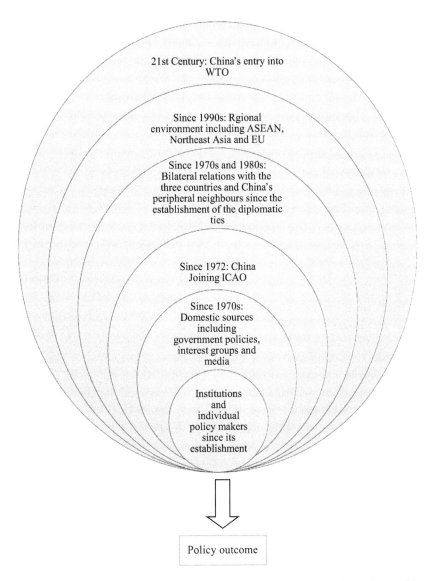

21st Century: China's entry into
WTO

Since 1990s: Rgional
environment including ASEAN,
Northeast Asia and EU

Since 1970s and 1980s:
Bilateral relations with the
three countries and China's
peripheral neighbours since the
establishment of the diplomatic
ties

Since 1972: China
Joining ICAO

Since 1970s:
Domestic sources
including
government policies,
interest groups and
media

Institutions
and
individual
policy makers
since its
establishment

Policy outcome

FIGURE 10.2 Factors and their evolution that have an impact on the policymaking
process.
Source: Compiled by the authors.

transport policy matters, in particular with respect to liberalisation and open
skies, there was little evidence that China's elites or very top leaders at the central
government level such as the Politburo or the SC have been noticeably involved
in the international air transport policymaking process, although they have been
the spearheads determining the strategic directions of China's overall foreign
trade policy. On some occasions, the State Councillors' instructions are sought,

but the day-to-day policymaking activities are managed by the bureaucrats at the ministerial level. This finding echoes Dougan's conclusion that the country's elites were not noticeably involved in matters with respect to civil aviation (Dougan, 2002).

Variations in achieving the policy goals across the examined country-pair markets

The extent to which China's international air transport policy can be considered as being liberal cannot be vested in a single "overnight" decision. As attested to, China has undergone a gradual process of identifying its regulatory policy position long after liberalisation had begun to take root in other advanced aviation economies globally. CAAC affirmed as much when it contended that its intent was to develop the international air transport industry in a "proactive, progressive, orderly and safeguarded" manner.

The true reflection of China's implementation of the above approach is thus vested in its bilateral negotiations with its international counterparts. This intent is manifested through the various negotiations and the concessions made by both parties concerned. Ultimately China's apprehension and restrictiveness diminished over this prolonged period as the country moves closer towards a liberalisation inflection point. In 1990, China was connected to the global community through just 44 international routes, by 2015 this had expanded to some 660. The year 2015 thus signalled a rapid increase in access with the number of international routes growing 120 percent over the quantum in 2010 (Figure 10.3). It is noteworthy that this expansion was not predicated on the restrictive bilateral policies.

The expansion was facilitated through the implementation of the CAAC's action plan to ensure aviation played a more central role in enabling national

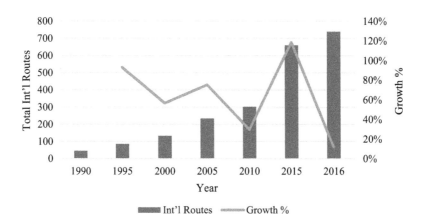

FIGURE 10.3 Expansion of China's international route access.
Source: CAAC Statistics Yearbook.

growth and development. By 2015, China had opened fifth freedom rights to 11 cities which improved the international networks of its airlines. There were reportedly 118 Air Service Agreements signed up to June 2016 according to state records. Of these, the majority were signed with other ICAO member states in Asia followed by Europe. On average, China has amended its ASAs with other Asian member states approximately 2.88 times each. This is in comparison with an average of 2.43 amendments per agreement in respect of its European counterparts. In total, there were 267 amendments, of which fifth to eighth freedom traffic rights were granted in 167 cases. The active rate at which China engaged counterparts to amend their ASAs signalled an increasing openness towards liberalising with all ASAs being amended an average of 2.2 times (Table 10.1).

Amendments to the original agreements were documented through various official documents and MOUs concluded at each round of negotiation. In this process, the 267 amendments China made signalled its increasing openness. A WTO system, which measures the openness of bilateral ASAs, suggests that up to 2016, China had only entered one "Type C" ASA (see Table 10.2 for detailed WTO ASA classification summaries). These allow third, fourth, and fifth freedom rights, and required double tariff approval, as indicated by the WTO. "Type C" ASAs are however constrained by operational limits requiring substantive ownership and effective control by nationals and regulatory pre-determinations of capacity. There was similarly 1 ASA regarded as Type E which primarily differs in relation to the multi-designation of carriers by the contracting states. The majority of the original agreements or 60 percent (71 of the 118) were restrictive Type A agreements which are typified as having characteristics including the third and fourth freedom rights, single designation, substantive ownership and effective control, double approval of tariffs, and capacity limitations.

The multiple amendments to the original ASAs as documented in MOUs are the true testimonials of China's appetite for incremental liberalisation. Accounting for these amendments, 41 (30 percent of the 118) of the existing agreements would now be considered to be "Type E," which highlights the movement

TABLE 10.1 Original ASAs and their amendments

Region	Total original ASAs	Total ASA amendments	Amendments per original agreement
Asia	43	124	2.88
Europe	37	90	2.43
South America	6	10	1.66
Africa	22	18	1.5
Oceania	6	7	1.16
North America	4	18	4.5
Total	118	267	2.26

Source: Various versions of CAAC Statistics Yearbook.

TABLE 10.2 Summary of WTO classifications for ASAs

ASA TYPE	Freedoms			Designation		O&C			Tariff structure			Capacity			
	3rd	*4th*	*5th*	*Single*	*Multi*	*SOEC*	*CI*	*PPB*	*Dbl Approv.*	*Free market*	*Dbl Disapp.*	*Pre-deter.*	*Free capacity*	*Bermuda I*	*Other*
A	X	X		X		X			X			X			
B	X	X			X	X			X			X			
C	x	X	X	X		X			X			X			
D	X	X	X	X		X			X						X
E	X	X	X		X	X			X			X			
F	X	X	X		X	X			X						X
G	X	X	X		X	X	X	X		X	X			X	
I	Incomplete as per ICAO codes														X
O	Any other combination														

Abbreviated terms are defined as follows: O&C, ownership and control; SOEC, substantial ownership and effective control; CI, community of interest; Pre-deter, predetermination of capacity.

Source: WTO (2011).

towards less restrictive policies than those embedded in "Type A" style agreements. The two core distinctions between "Type A" and "Type E" style agreements are designation of carriers on routes and the award of fifth freedom rights. Whereas Type A agreements permit signatory countries to each designate one airline on the route, Type E agreements allow for multiple designations. Similarly, Type A agreements do not provide fifth freedoms whereas this is encompassed in Type E agreements. Additionally, the amendments have resulted in some 11 ASAs, which now have characteristics that would reflect "Type G" agreements, signifying ASAs in liberal forms. These agreements are typified by characteristics such as the third, fourth, and fifth freedom rights and multi-designation. Unlike other ASAs which seek to enforce nationality clauses, Type G ASAs provide for alternative forms of ownership and control and also consider operations with the principal place of business in the contracting party of a community of interest designation. There are also no controls on pricing and capacity.

In China's case, as it began to gradually liberalise its markets through negotiations of amendments of its ASAs, the majority of the amendments held by the CAAC's records up to 2016 were to amend capacity constraints. These increases thus provided greater prospects to respond to buoyant market demand in the respective cities. Similarly, substantive amendments were also made to integrate new city pairs which facilitated and gave the carriers the ability to strengthen their market presence and dominance on the city pairs. Considering the integration of the amendments, China's most liberal agreements are vested in its ASAs with North American, ASEAN, and European member states in keeping with their strong trade ties.

China's propensity towards a gradual and progressive opening is thus aptly portrayed through its staggered and consistent negotiations which incrementally dissolve its apprehension to liberalising with counterparts. Charting the negotiations of the case study countries elucidates this. Thus far, it has not been proven to be the Chinese style to liberalise in "one fell swoop." Even in this instance when liberalisation was pursued in the domestic market in tandem with the overall economic reforms, it was staggered. Notwithstanding pressures from liberal counterparts like the USA, China maintained a steady resolution and commitment to being progressive and staggered in its openness engaging in multiple consultations with its counterparts. In the case of the USA, there were some five consultations between 1992 and 2007 (see Figure 10.4).

Critics of globalisation and trade liberalisation contend that the developed world showed affinity for general liberal trade policies only after their domestic markets become competitive enough to withstand the pressures of the imported competition which liberalisation would bring. In this regard, it is argued that trade liberalisation was moderated in some sense by the level of domestic appetite and readiness for external competition. In the case of aviation, there was no need to delay liberalisation on account of securing the domestic market. With the entrenched regulatory structure preserving the integrity of the domestic market, there existed no real threat to domestic competition and carriers operating within

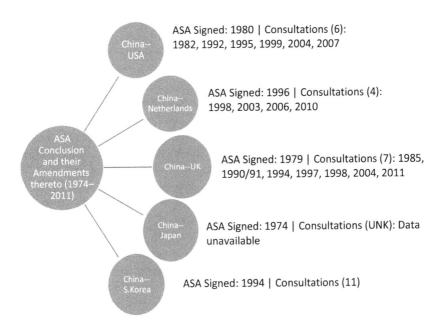

FIGURE 10.4 Summary of negotiations with case studies of counterpart nations.
Source: Compiled by the authors.

the domestic space. China's case is somewhat more peculiar at the domestic level and is worth mentioning in the context of its staggered approach to relaxing the constraints as contained in its ASAs. As discussed in Chapter 9, there was a dismantling of the socialist institutions and the evolution of the market-oriented drivers as a part of the economic reforms. What is critical is that the move from the centrally controlled economy to a market economy increased the prevalence and involvement of non-state actors. There were now competing interests which the CAAC had to balance that did not exist in the same context earlier.

Competition amongst domestic carriers has an impact on international performance by the airline (Clougherty and Zhang 2009). On this basis, the lack of domestic competition as perpetuated through the state control over the industry would have had an effect on the international competitiveness of Chinese carriers. Since the late 1990s, the corporatised SOE airlines began to unbundle their close intimate relationships with the CAAC developing their own strategic visions. It was similarly during this period that China began to exert greater propensity towards negotiations that opened its ASAs. China's move to a market economy meant that there were divergent non-state opinions to be harmonised. As was further discussed in Chapter 9, CAAC was often caught in between these divergent opinions. With this in mind, the evolving nature of China's domestic market and influences were either consequential or necessitated this staggered approach to liberalisation. Without it, there were greater risks of alienating or adopting an approach which overly appeased select elements of the burgeoning domestic influences and interests.

Originality and contribution to knowledge

Original research is both the creation of new knowledge and an innovative approach to the discovery of that knowledge (Liley and Mullin, 2005). Originality in research does not necessarily mean an invention out of a vacuum but can be the gathering-together of the known elements that hitherto have been kept apart (Finn, 2005). Through a systematic, synthesised, and critical analysis of the known data, a researcher is able to uncover new knowledge either via the discovery of new facts and evidence or via the innovative re-interpretation of known data and established ideas (Murray, 2002), thus introducing new concepts and theories that are regarded as novel, creative and innovative that contribute to knowledge.

This empirical research has been undertaken at the inspirations of the authors, who aspired to unwrap the myths of China's international air transport policymaking, which is hardly tapped in the existing literature. Treating international air transport as an international trade in services, the authors selected several country-pair markets as case studies. Applying the Micro-Macro Linkage Approach, the authors intended to identify factors from various sources, which include but are not limited to international regimes and institutions, regional integration, domestic politics, interest groups, organisational structure, and bureaucrats, that have had an impact on the country's international air transport policymaking process. The authors also set out to establish how these factors have interacted with each other and converged on the policymakers, who have ultimately crafted the policy direction and shaped the policy outcome. The study showcases how China's political and economic reform and the global environment have affected the country's political system. Executing the research has aided in establishing contributions to knowledge varying from empirical insight about china's air transport policy to the application of that empirical data to test the Macro-Micro Linkage Model.

Novelty of undertaking international air transport policymaking from political economic perspective, and examining the policy outcome through the negotiations

This research is the first detailed attempt to assess air transport industry as a case study within the domain of international trade policy. In so doing, it presents readers with a novel research design when compared to other studies and works in the field and on this basis presents an innovative inquisition for readers in both air transport and international trade disciplines. While other studies have examined the implications of liberalisation (i.e. airline competitions, traffic growth, and consumer benefits), there have been no works which have undertaken empirical research in international air transport from this type of disciplinary perspective. As with any multidisciplinary design, the core advantage of adopting this approach is that it has enabled theoretical considerations which could only have emerged through the study and application of idiosyncrasies associated with the realm, concepts, ideas and ideologies vested within the domains of trade theory and political economy scholarship.

Approaching cross-border international air transport policymaking and its execution from bilateral negotiation perspective

This synthesis is uniquely situated within a new domain due to its novel application of trade policy scholarship in China's air transport policy. The arduous efforts in chronicling the negotiations between China and other counterparts advance knowledge about how China approaches and formulates its trade policy considering its peculiarities. Having performed this analysis and formulated a discussion in this context, readers in the fields of trade policy can comprehensively orient themselves to a study of this discipline in the context of China and other bilateral partners. For readers in aviation, it provides a more comprehensive understanding of the negotiations within the broader context that is the nexus between China responding to its domestic circumstances whilst attempting to appease its international obligations.

The establishment of the GATS in 1995 created an intergovernmental agreement with a framework of agreed principles. This is one of the core distinctions between international trade in air transport and other services industries. Whereas liberalisation in other sectors such as finance is largely pursued without the need for bilateral discourse, liberalisation in air transport largely remains a bilateral activity, despite some multilateral arrangements being achieved. Consequently, viewed from a trade policymaking and its execution perspective, thrusts towards liberalism in international air transport arguably remain solely within the domain of a nation state. In this respect, while the cross-border air transport policy maybe declared as being liberal, the extent and scope of such liberalism is achieved through various rounds of bilateral negotiations. When taken collectively, these negotiations form the member states' overall trade policy in international air transport.

More empirical data has been collected revealing new insight

An enormous amount of secondary data from historical files and records and invaluable primary data collected through dozens of interviews has enabled the authors to present a more rigorous and accurate exposition in the context of this work. The empirical value derived from this effort can be found in its combination of relevant information into one source, making knowledge about China's international air transport policymaking more readily available for further research. These findings as well as the discussions, analysis and conclusions as presented in preceding chapters will hopefully help to improve understanding about China, its aviation industry, policymaking process, and the formulation of its overall response to positions and mandates established by the international community.

Applying the Macro-Micro Linkage Model in a new field of study, and testing its vigorousness

The Micro-Macro Linkage Model was developed by Zhao (1996) to understand how variables at both the micro and macro levels have an impact on China's policymakers, and how the identified factors have shaped and crafted their mindset and perception, resulting in a particular policy choice. He applied his model to analyse the evolution of China-Japan relationship between 1979 and 1995 and suggested a very comprehensive agenda for further research, including case studies on bilateral relations between China and its counterparts on its key policy decisions.

Zhao (1996) expressed his profound hope that more research needs to be conducted leading to empirical findings at the micro level so as to enable a full coverage of the interactions between factors at different levels with regard to foreign policy choices, hence contributing to a theoretical framework (Zhao, 1996). The last two decades have seen an increasing number of research on China's foreign policymaking, including foreign trade policymaking, with an overwhelming focus on the influence of domestic factors. Yet, no further empirical evidence has been collected to test this Model. This is likely the result of several considerations. Firstly, foreign trade policymaking sits in an interdisciplinary area. Secondly, there exist several limitations about the Model, which Zhao acknowledges. The model was intended to identify various sources of factors at different levels which interact with each other with an impact on the policy outcome. "The limit of the study is the insufficiency of empirical evidence at micro level, which would allow a comprehensive application of this Model, to capture the interactions between factors at various levels with regard to foreign policy choices" (Zhao, 1996). He further acknowledges that "the major challenge to furthering the Linkage Approach is to conceptualise a theory, which could be used to translate and link the factors at the individual level to the factors characterising social systems, and vice versa" (Zhao, 1996).

While this clarion call had largely been neglected, it has been attended to by this research, in spite of the fact that Zhao suggested a very comprehensive agenda for further research. Empirical evidence is produced at both the macro and micro levels for the country-pair markets. At the macro level, the relationships between international organisations and China and between individual countries have been discussed. More importantly, as urged by Zhao, at the micro level, domestic sources such as societal factors, institutions, individual decision makers, as well as their relationships with each other have also been examined.

By applying the Micro-Macro Linkage Model to the industry cases with a longitudinal approach, causal relationships between the variables and the policy outcome can be proposed. Applying the model also assists in establishing if there is any pattern in terms of policymaking and policy change. Although the research did not lead to any substantial theoretical formation, it is a significant effort in testing the vigorousness of the Model in different settings. It has produced a substantial amount of empirical evidence demonstrating the relevance of the Model, which can be developed to generate theories more applicable to analysing China's foreign trade policymaking. From this perspective, this research is a meaningful effort at applying the Model to new areas.

A revised model to analyse China's international air transport policymaking

International air transport is by nature a commercial activity in the realm of international trade. Although the political significance attached to air transport has been fading away since the US launched the open skies in 1992, primary considerations for liberalising international air transport market are given to principles of sovereignty, opportunity of equality and fairness. Negotiations of cross-border market entry are conducted at the governmental level, with national interest having remained the priority consideration.

Empirical evidence collected from this process and the preceding analysis has established the need to develop a revised model in order to allow any researchers interested in this subject matter to pursue further. Figure 10.5 depicts the revised model and the description of each element.

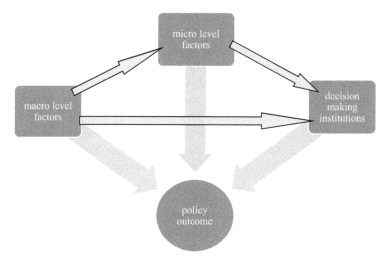

FIGURE 10.5 Revised Macro–Micro Linkage Model.
Source: Compiled by the authors.

The macro-level factors refer to the environment at international, regional and bilateral levels. They include international regime, international law, international organisations, regional cooperation bodies, political and economic forums, and bilateral relations that are out of control of the sovereign state. They have a direct impact on micro-level factors as well as on individual decision makers. Similarly, micro-level factors indicate the sources at the domestic level which include the state government, its national policy, interest groups, media, public opinions, think tanks and institutions that are operating within the boundary of a nation state. They, too, have a direct impact on the individual decision makers and the policy outcome. The revised Model emphasises a three-way interaction of those factors identified at each level. While constraints at the international level and determinants at domestic level are involved in a dynamic relationship at the macro level, they also converge on, and receive feedback from, individual decision makers. Factors at these levels together exert an influence on the final policy outcome, though the weight of the factors at each level might be different.

Limitations of the research

While this research has made considerable contribution to knowledge, we acknowledge the following limitations.

- Firstly, the sample size could be expanded to include more country-pair markets as case studies. For example, a couple of African countries could be selected to analyse how China has negotiated its bilateral air service agreement and China's approach towards liberalising the country-pair markets. This inclusion would enable a more comprehensive comparison on China's approach towards liberalisation;

- Secondly, there was a limited access to secondary data. This was a result of filing policies and procedures in place, which was out of the control of the researchers. Should more secondary data from governments and corporate sources be available, in particular, the files and records relevant to some key country-pair markets such as Japan, South Korea, the USA, and the EU countries, a more thorough and comprehensive account and analysis could have been undertaken to examine the latest development of China's approach towards liberalising the bilateral air transport market and how the country has negotiated and achieved its policy objectives. The limited access to files concerning China–Japan and China–South Korea markets only allowed the researchers to investigate the two cases to a certain extent, which has thus adversely affected the in-depth analysis of the subject matter;
- Thirdly, analysis has not been undertaken to assess the economic impact resulting from each round of negotiations between the selected country-pair markets. This is determined by the scope of this research, taking into account the time and financial resources available.
- Finally, there is some constraint about the Micro-Macro Linkage Model itself, which was originally designed to analyse China's foreign policymaking. The Model was nevertheless revised and should now be further tested in the same sector or other industries to further knowledge advancement.

Suggestions for future research

The following has been identified for further studies:

- Firstly, as China deepens its economic reform and increasingly integrates itself into the global economic system, it has transformed from a planned economy to a market economy. It is more exposed to various sources of influence from an increasing number of factors and driving forces. Identifying those factors and driving forces is an evolving task and a holistic and longitudinal approach is essential and appropriate. More research is needed to establish the factors and driving forces in the past decade since President Xi took office, and to determine whether and to what extent these driving forces have been evolving.
- Secondly, bilateral negotiations have brought about significant changes in procedures for developing international air transport policy, involving a growing number of public and private actors in the process. What are the negotiation styles of China's international air transport policymakers and how is the policy outcome affected through the negotiation process?

- Thirdly, while corporate political activities are common in the western business community, which are also observant in the Chinese context, some empirical evidence needs to be established. In the air transport sector, are aviation entities in China, such as airlines, airports, aviation supply businesses, joint ventured airlines, airline alliances, aircraft manufacturers, and associations different from businesses of other industries when undertaking such activities? If different, how different are they in terms of behaviour and influencing power?

- Fourthly, the CAAC, as the sole industry regulator and policymaker with respect to air transport, has not been involved too heavily in bureaucratic bargaining in the determination of the nation's international air transport policy, although it needs to consult with other government agencies regularly and constantly. The CAAC also oversees the air transport industry vertically across the country without much interference from local governments and other bureaucratic institutions. This is in contrast to other Chinese ministries whose authority is fragmented with tens of other bureaus being involved in the process of coordination before seeking a consensus and with local governments who usually have an economic stake in the business. Consequently, does this kind of vertical oversight of the industry shape the CAAC's approach to policymaking compared with other industries? If the answer is yes, then how differently does the CAAC behave when compared with other agencies? What are the features of the policymaking process of both CAAC and other ministries, and to what extent do these different features have an impact on the policymaking process and policy outcome? To what extent do these kinds of institutional features influence the nation's policymaking? To what extent does this have an impact on the negotiations with counterparts and to what extent can key individual negotiators manipulate the situation to achieve the best policy choice?

- Another area worth exploring further is the impact on individual policymakers of the changing personal mindset. As growing numbers of Chinese bureaucrats have overseas experience, it is claimed in some research that personal experience has played a positive role in shaping a person's ideology and gradually changing their mindset. Air transport is an internationally standardised business with its employees having ready exposure to global influence. Pilots as well as other professions in the air transport industry already have a good knowledge of the international standards. Does this exposure have an impact on the policymakers' mindset, thus affecting the policymaking process? To what extent does this overseas experience, standardised professional background and international exposure have an impact on the policy choice in terms of international air transport, especially on a country-pair market?

- Finally, a mechanism is necessary so as to allow any researcher to measure the identified factors and further determine their weighting in the policymaking process. This will help to address the concerns raised by Kim (1994), who warns that a list of laundry-list-like factors should be avoided when assessing the influence of the identified variables. This can be very challenging, though, as the variables established out of each empirical research can be divergent, thus causing difficulty in getting to an agreement as to what can be defined as a causal factor and what can be a consequence.

Policy implications and recommendations to the government and industry

It was premised in Chapter 1 that formulating an international air transport policy is a binary consideration of domestic economic circumstances and international issues. International air transport policy is thus not established in isolation and is evidently a fragment of a nation's overarching trade policy. Under the parameters of the WTO, trade is meant to absolve discriminatory and preferential practices and ensure fair market access. Similar parameters are not substantively entrenched in international air transport. Evidently, to the extent that the economic and commercial traffic rights have been subject to bilateral negotiation in air transport, there has generally been a greater propensity for nations to use air transport as a medium through which to exert leverage in their overarching national trade policy and even in relation to their geopolitical objectives. For example, concerted efforts by the USA to liberalise with China's peripheral neighbours exert a similar liberalisation force on China. While it is true that China had been slow to open and liberalise in comparison with other major air transport markets, the CAAC's dominance for decades as the agency of record for administering China's international air transport policy has enabled it to develop a cohesive policy stance which fulfils the respective national objectives as set by the central government.

The structure of China's political economy has been typified by waves of reform which have encouraged its air transport policy to be non-fragmented, gradual, and iterative. For the most part, China's international air transport policy has been cohesive across the counterparts examined. This has emerged from the autonomy which the CAAC has enabling it to ensure a cohesive policy. With respect to trade in other services, such as financial services, cohesiveness in trade policy would be maintained through the dictates of the country's international obligations. A counterpart nation would reasonably expect fairness and equity in market access, and through supranationally agreed dispute mechanisms, can appoint onto itself the right to challenge a member state's policy where that counterpart nation suspects it is being disadvantaged by preferential or discriminatory treatment. For example, in 2012, Panama lodged a complaint with the WTO asserting that provisions established in Argentina's domestic law targeted at services and providers from jurisdictions which did not exchange information with Argentina was discriminatory. *Inter alia*, the measure imposed additional financing, registration, and taxation requirements. The essence of this reference in relation to the discourse herein is typified through the recourse available to the aggrieved nation which perceived its rights to fair and equitable market access were infringed.

Without the adoption of a supranational agreement which facilitates a principle akin to that of the most favoured nation principle in international trade which enshrines practices of fair trade, nation states also afford onto themselves the liberty to determine the extent of their accessibility in their international air transport policy. Even in a context where international air transport policy is fashioned on the basis of reciprocity, it still remains within the state's domain to determine the scope of its liberalism even if the bilateral partner is committed to absolving all restrictions to its own market. This is often reflected in traffic limits applied in key/hub cities.

In the context of the negotiations examined, China's approach and the autonomy of the CAAC serve as internal barometers to preserve the integrity of its internal air transport policy position regardless of the counterpart nation. From China's perspective, this could enhance its ability to ensure that it does not bind itself to an international air transport agreement in relation to one nation that is less favourable than others because of vacillations in prior policy positions driven by excessive extraneous influences. When there is interference or decision-making is too pluralistic, equality is undermined. For example, where international air transport serves as a trade-off towards another diplomatic end, this does not mean that *ceteris paribus*, the same outcome could be achieved.

International aviation policy in China has thus far principally developed in keeping with how the CAAC interpreted its mandate to enable the national development objectives as set forth by the central government. This is important for understanding how the commercial policy adopted by the central government can be executed cohesively when channelled through a single entity. Trade policy in aviation is applied conversely to other industries since it is applied at the bilateral level. In cases like this, the less pluralistic the decision-making, the more cohesive the application of the trade policy.

Decision-making is an art rather than a science. Process affects substance (Cohen et al., 2003). By identifying the factors that have an influence on the policymakers, and examine how these factors have affected the policy goals, we attempt to unwrap the black box of China's international air transport policymaking and approach to liberalising its international air transport market. The following recommendations are made to the government and industry:

• The policymaking process needs to be systemised and institutionalised. A set procedure needs to be developed with clear guidelines available for stakeholders to follow. The set procedures would help to avoid any arbitrary decisions;

- Lobbying channels need to be formalised to allow various views and opinions to be lodged. The formalised channels would provide a platform for debate for the best possible policy outcome;
- Industry and stakeholders should approach policymakers through official channels. It would avoid the situation where voices are conveyed through personal contacts which might produce biased personal preferences;
- In presenting the case, industry and stakeholders should base their arguments more on objective and comprehensive market research to best reflect their interests rather than be simply responding to the central government's call. This could avoid the unnecessary waste of resources.

Challenges ahead

It has been more than 30 years since the USA first deregulated its domestic air transport market. The spill-over effects have spread rapidly, resulting in other countries around the world following suit. Liberalising the international air transport market has been a major theme over the past two decades, with the EU–US open skies arrangements being a milestone in lifting the economic constraints of airline operations across borders. Regardless, open skies agreements still only go half way to normalising the industry, as they continue to restrict access to domestic markets and leave ownership and control rules untouched (UK CAA, 2006). Hence, creating a single aviation market[1] becomes an ultimate objective of policymakers so that not only restrictions on trade barriers such as market entry and capacity are removed but also the limitations on control and ownership of national airlines are relaxed, which will allow airlines to optimise the opportunities in global capital markets and to have access to domestic markets in other countries.

The primary rationale for open skies is to optimise the economic achievements (Forsyth et al., 2004). When implemented effectively, open skies will produce comprehensive net economic benefits for the countries participating in the arrangement (Forsyth et al., 2004). Liberalisation of the regional air transport market could have an enormous and positive impact on regional prosperity (InterVISTAS, 2006). Liberalising a domestic market means overall economic benefits for an individual country, such as generating more job opportunities and contributing to their GDP growth. Airlines would be more flexible to design and develop their route networks as they intend, offering travellers more choices for flights with lower fares.

However, liberalisation could also have an adverse effect on the airline industry, which might result in long-established national carriers going bankrupt. National governments are equally concerned about the non-economic consequences such as national defence and security when considering the liberalisation policy (Forsyth et al., 2004). While national governments are increasingly willing to relax economic controls over airline operations, they are less willing to agree to anything that has an implication to cause the surrender, or even partial surrender, of their sovereignty in the matter of traffic rights (Hanlon, 2007), which is at the core of Chicago Convention. Caught in this paradox, individual governments could be reluctant to make a quick move towards full open skies. Agreements become so challenging to achieve, especially between nations that are not like-minded or share the belief that more open markets would serve consumers better and improve the prosperity of society as a whole, regardless of the performance of individual airlines.

Notwithstanding this, the arguments for and against liberalisation have never ceased. Those who have voiced for full and complete liberalisation are able to cite numerous empirical evidences on the impact of liberalisation, highlighting the benefits for industry, consumers, and society. Those who have argued for re-regulation are mainly concerned about the safety implications and the poor financial performance of the industry, citing the inconsistency of various regulatory regimes that might lead to free-riders, thus raising strong commercial, strategic, and presentational concerns about the fairness of allowing ownership of domestic airlines by nationals from countries which do not have broadly equivalent open investment rules (UK CAA, 2006). One of the key challenges is how the complex framework of interdependent bilaterals might be unwound while maintaining effective regulatory control over safety (UK CAA, 2006). Another challenge is how to address the issue of regulatory divergence, namely, the distortionary impact of different regulatory approaches applying to competitors operating in the same market. To this effect, a regulatory convergence is required to accommodate the differences so as to allow fair competition (UK CAA, 2006).

In the past few decades, China's air transport industry has experienced an exponential growth, at an average of 9 percent per annum. Ironically, this phenomenal achievement has been made in the course of the country's regulatory reform, which is still undergoing a liberalisation process, both domestically and internationally. With its huge population and latent demand, the Chinese market is set to become one of the biggest in the years to come (Hanlon, 2007). With liberalisation being an irreversible trend all over the world, China will no doubt follow this course. Its approach to further liberalising its domestic and international air transport market, its policy directions, measures to take to execute the policy, and its manners in dealing with its counterparts in this domain will be crucial for its long-term prosperity. More importantly it will affect its peripheral neighbours in the region, its trade partners, as well as others who might not have a stake in China. It is left to be seen what global impact will result from lateralization thrusts in China given its ever expanding role in the global community.

Note

1 A Single Aviation Market already exists within the 27 member states of the EU as a result of the creation of a European Single Market and between Australia and New Zealand.

References

Clougherty J. A. and Zhang, A., (2009), Domestic rivalry and export performance: theory and evidence from international airline markets. *Canadian Journal of Economics*, 42(2), 440–468.

Cohen S. D., Blecker R. A., and Whiney P. D., (2003) Fundamentals of US foreign trade policy, economics, politics, laws and issues; (2nd edition), Westview Press, Boulder, CO and Oxford.

Dougan M., (2002), *A political economy analysis of China's civil aviation industry*, Routledge, London.

Finn J. A., (2005), *Getting a PhD: an action plan to help manage your research, your supervisor and your project*, Routledge, Abingdon.

Forsyth P., King J., Rodolfo C. L., and Trace K., (2004), Preparing ASEAN for Open Sky, AADCP regional economic policy support facility research project 02/008, final report, available at www.asean.org/aadcp/repsf/docs/02-008-ExecutiveSummary.pdf

Hanlon P., (2007), *Global airlines, competition in a transnational industry*, (3rd edition), Butterworth Heinemann, Elsevier Ltd, Oxford.

ICAO, (2010), Overview of trends and developments in international air transport, Montreal, available at https://www.icao.int/sustainability/Documents/OverviewTrends.pdf

ICAO, (2013), Worldwide air transport conference sixth meeting: liberalization of market access, available at www.icao.int/Meetings/atconf6/Documents/WorkingPapers/ATConf6-wp060_en.pdf

InterVISTAS-ga2, (2006), *The economic impact of air service liberalisation*, Bethesda, the USA.

Kim S., (1994), China's international organisational behaviour, in: Robinson T. W., and Shambaugh D., (editors), *Chinese foreign policy, theory and practice*, P405, (pp. 401–434), Oxford University Press Inc., New York

Kong B., (2010), *China's international petroleum policy*, Greenwood publishing group, Santa Barbara, CA.

Lai H., (2010), *The domestic sources of China's foreign policy, regimes, leadership, priorities and process*, Routledge, Abingdon.

Lieberthal K., and Oksenberg M., (1988), *Policy making in China: leaders, structures, and processes*, Princeton University Press, Princeton, NJ and Oxford.

Liley M., and Mullins, G., (2005), Supervisors' conceptions of research: what are they? *Scandinavian Journal of Educational Research*, 49(3), 245–262.

Murray, R., (2002), *How to write a thesis*. Open University Press, Berkshire.

Simmons B. A., and Martin L. L., (2002), International organizations and institutions, in: Carlsnaes W., and Simmons B. A., (editors), *Handbook of international relations*, (pp. 326–351), SAGE Publications Ltd., London and California.

Snedden B., (1981), Foreword, in: Poulton H. W., (editors), *Law, history and politics of the Australian two airline system*, p. xv, Parkville, Melbourne.

UK CAA, (2006), Ownership and control liberalization, a discussion paper, CAP769, The UK Civil Aviation Authority, London.

WTO, (2011), *Services: transport — air transport: air services agreements projector* (ASAP), World Trade Organisation, Geneva.

Zhao Q., (1996), *Interpreting Chinese foreign policy, the micro-macro linkage approach*, Oxford University Press (China) Ltd., Hong Kong.

INDEX

Note: **Bold** page numbers refer to tables and *italic* page numbers refer to figures.

For Product Safety Concerns and Information please contact our EU
representative GPSR@taylorandfrancis.com
Taylor & Francis Verlag GmbH, Kaufingerstraße 24, 80331 München, Germany